Lecture Notes in Bioinformatics   7348

Edited by S. Istrail, P. Pevzner, and M. Waterman

Subseries of Lecture Notes in Computer Science

W0246089

Olivier Bodenreider   Bastien Rance (Eds.)

Data Integration in the Life Sciences

8th International Conference, DILS 2012
College Park, MD, USA, June 28-29, 2012
Proceedings

 Springer

Series Editors

Sorin Istrail, Brown University, Providence, RI, USA
Pavel Pevzner, University of California, San Diego, CA, USA
Michael Waterman, University of Southern California, Los Angeles, CA, USA

Volume Editors

Olivier Bodenreider
Bastien Rance
National Institutes of Health, US National Library of Medicine
8600 Rockville Pike, Bethesda, MD 20894, USA
E-mail: {olivier.bodenreider, bastien.rance}@nih.gov

ISSN 0302-9743   e-ISSN 1611-3349
ISBN 978-3-642-31039-3   e-ISBN 978-3-642-31040-9
DOI 10.1007/978-3-642-31040-9
Springer Heidelberg Dordrecht London New York

Library of Congress Control Number: 2012939199

CR Subject Classification (1998): H.3, J.3, I.2, H.4, H.2, H.5

LNCS Sublibrary: SL 8 – Bioinformatics

Typesetting: Camera-ready by author, data conversion by Scientific Publishing Services, Chennai, India

Printed on acid-free paper

Springer is part of Springer Science+Business Media (www.springer.com)

Preface

This volume of *Lecture Notes in Bioinformatics* (LNBI) contains selected papers from the 8th International Conference on Data Integration in the Life Sciences (DILS 2012), held June 28–29, 2012 at the University of Maryland in College Park, Maryland, USA (http://sites.google.com/site/webdils2012/).

The Data Integration in the Life Sciences (DILS) conference has been held regularly since 2004, alternating between venues in North America and Europe. Over the years, DILS has become a forum for life science researchers, a place where issues in data integration are discussed, where new avenues are explored, and where integration is extended to new domains. Through a mix of invited keynote presentations, oral presentations of peer-reviewed papers, posters and demos, a variety of ideas are discussed, ranging from reports on mature research and established systems, to exciting new prototypes and ongoing research.

This year the conference was organized around three major themes. Each session was introduced by a keynote presentation followed by paper presentations. In the "Foundations of Data Integration," Jim Ostell from the National Center for Biotechnology Information (NCBI) presented the Entrez system. Recurring themes from the papers included ontologies, semantic similarity, mapping between ontologies and schema matching. The second theme, "New Paradigms for Data Integration," was introduced by a presentation of the Watson system by Ken Barker from IBM Research. The papers demonstrated the benefits of Semantic Web technologies for integrating biological data and explored crowd-sourcing as a potential resource to support data integration. Finally, DILS 2012 emphasized "Integrating Clinical Data." Jim Cimino from the National Institutes of Health (NIH) Clinical Center presented "BTRIS," the clinical data warehouse supporting translational research at NIH. The papers explored the integration of clinical data for cancer research and the contribution of natural language processing to the integration of unstructured clinical data.

We thank the Program Committee for thoroughly reviewing and helping to select the manuscripts submitted to the conference. Our thanks also go to Louiqa Raschid, who coordinated the logistics at the University of Maryland.

June 2012
Olivier Bodenreider
Bastien Rance

Organization

DILS Steering Commitee

Sarah Cohen-Boulakia	LRI, University of Paris-Sud 11, France
Graham Kemp	Chalmers University of Technology, Sweden
Patrick Lambrix	Linköping University, Sweden
Ulf Leser	Humboldt-Universität zu Berlin, Germany
Bertram Ludaescher	University of California, USA
Paolo Missier	Newcastle University, UK
Norman Paton	University of Manchester, UK
Louiqa Raschid	University of Maryland, USA
Erhard Rahm	University of Leipzig, Germany

General Chairs

Olivier Bodenreider	National Library of Medicine, NIH, USA
Louiqa Raschid	University of Maryland, USA

Webmaster

Bastien Rance	National Library of Medicine, NIH, USA

Local Organizers (USA)

Louiqa Raschid	University of Maryland, USA

Program Committee

Christopher Baker	UNB Saint John Faculty, Canada
Elmer Bernstam	University of Texas, USA
Judy Blake	The Jackson Laboratory, USA
Anita Burgun	Université de Rennes 1, France
Jim Cimino	National Library of Medicine, NIH, USA
Sarah Cohen-Boulakia	LRI, University of Paris-Sud 11, France
David De Roure	Oxford e-Research Center, UK
Dina Demner-Fushman	National Library of Medicine, NIH, USA
Michel Dumontier	Carleton University, Canada
Christine Froidevaux	LRI, University of Paris-Sud 11, France
Carole Goble	University of Manchester, UK
Graciela Gonzalez	Arizona State University, USA
Vasant Honavar	Iowa State University, USA

Tony Xiaohua Hu	Drexel University, USA
Hasan Jamil	Wayne State University, USA
Graham Kemp	Chalmers University of Technology, Sweden
Purvesh Khatri	Stanford Universtiy, USA
Patrick Lambrix	Linköping University, Sweden
Adam Lee	National Library of Medicine, NIH, USA
Mong Li Lee	National University of Singapore, Singapore
Ulf Leser	Humboldt-Universität zu Berlin, Germany
Frédérique Lisacek	Swiss Institute of Bioinformatics, Switzerland
Bertram Ludaescher	University of California, USA
Yves Lussier	University of Chicago, USA
Brad Malin	Vanderbilt University, USA
M. Scott Marshall	University of Amsterdam, The Netherlands
Marco Masseroli	Politecnico di Milano, Italy
Paolo Missier	Newcastle University, UK
Peter Mork	MITRE, USA
Fleur Mougin	University of Bordeaux 2, France
Shawn Murphy	Partners HealthCare, Boston, USA
Radhakrishnan Nagaraja	University of Arkansas for Medical Sciences, USA
Jyotishman Pathak	Mayo Clinic College of Medicine, USA
Norman Paton	University of Manchester, UK
Erhard Rahm	University of Leipzig, Germany
Bastien Rance	National Library of Medicine, NIH, USA
Dietrich Rebholz-Schuhmann	EBI, UK
Alan Ruttenberg	University at Buffalo, USA
Satya Sahoo	Case Western Reserve University, USA
Neil Sarkar	University of Vermont, USA
Guergana Savova	Children's Hospital, Boston, USA
Michael Schroeder	TU Dresden, Germany
Nigam Shah	Stanford University, USA
Amit Sheth	Wright State University, USA
Andrea Splendiani	Rothamsted Research, UK
Karin Verspoor	National ICT, Australia
Maria Esther Vidal	Universidad Simòn Bolìvar, Venezuela
Chris Welty	IBM, USA
Guo-Qiang Zhang	Case Western Reserve University, USA

Additional Referees

Artjom Klein	UNB Saint John Faculty, Canada
Thomas Wächter	TU Dresden, Germany
Robert Leaman	Arizona State University, USA
Daniel Eisinger	TU Dresden, Germany
Anne Morgat	Swiss Institute of Bioinformatics, Switzerland
Alexandre Riazanov	UNB Saint John Faculty, Canada

Table of Contents

Foundations of Data Integration

Entrez: The NCBI Search and Discovery Engine 1
James M. Ostell

InterOnto – Ranking Inter-Ontology Links 5
Silke Trißl, Philipp Hussels, and Ulf Leser

Finding Cross Genome Patterns in Annotation Graphs 21
Joseph Benik, Caren Chang, Louiqa Raschid, Maria-Esther Vidal,
Guillermo Palma, and Andreas Thor

Pay-as-You-Go Ranking of Schema Mappings Using Query Logs 37
Ruhaila Maskat, Norman W. Paton, and Suzanne M. Embury

New Paradigms for Data Integration

Combining Structured and Unstructured Knowledge Sources
for Question Answering in Watson 53
Ken Barker

Cancer Data Integration and Querying with GeneTegra 56
E. Patrick Shironoshita, Yves R. Jean-Mary, Ray M. Bradley,
Patricia Buendia, and Mansur R. Kabuka

Integrating Large, Disparate Biomedical Ontologies to Boost Organ
Development Network Connectivity 71
Chimezie Ogbuji and Rong Xu

Validating Candidate Gene-Mutation Relations in MEDLINE Abstracts
via Crowdsourcing 83
John D. Burger, Emily Doughty, Sam Bayer, David Tresner-Kirsch,
Ben Wellner, John Aberdeen, Kyungjoon Lee, Maricel G. Kann, and
Lynette Hirschman

Integrating Clinical Data

The Biomedical Translational Research Information System:
Clinical Data Integration at the National Institutes of Health 92
James J. Cimino

ONCO-i2b2: Improve Patients Selection through Case-Based
Information Retrieval Techniques 93
      Daniele Segagni, Matteo Gabetta, Valentina Tibollo,
      Alberto Zambelli, Silvia G. Priori, and Riccardo Bellazzi

Syntactic-Semantic Frames for Clinical Cohort Identification Queries ... 100
      Dina Demner-Fushman and Swapna Abhyankar

Author Index 113

Entrez: The NCBI Search and Discovery Engine

James M. Ostell

National Center for Biotechnology Information, NIH, Bethesda, MD
ostell@ncbi.nlm.nih.gov

Abstract. About 2.7 million unique users a day come to the NCBI web-
site to explore data resources as diverse as the PubMed database of published scientific articles, the PubMed Central (PMC) database of full text
articles and books, the GenBank database of DNA and protein sequences,
the dbSNP database of genetic variation, or the PubChem database of
chemical structures. They search and retrieve records in each very different database, but also follow the connections between them such as the
article in PubMed where the DNA sequence in GenBank was reported,
or the protein that would be altered by a variation in dbSNP. While the
names of the individual databases are household words in biomedicine
and biomedical researchers just assume they have ready access to them
all and can easily navigate through the relationships among the database
as a matter of course, a great deal of design, engineering, and software
development has gone into making this possible. Underlying the search
and retrieval of these databases is the Entrez search and retrieval engine.

Keywords: NCBI, PubMed, GenBank, Entrez.

1 Introduction

Entrez is the search and retrieval system that supports both interactive Web-
based searching and display, as well as an application programming interface
(API), Eutils, to the same content and capabilities, which can be integrated into
outside tools and websites. Entrez supports a significant fraction of interactive
web use at NCBI, producing about 12 million web page views a day for about 1.6
million unique users a day, or about 1.2 terabytes of output a day. At peak rates,
Entrez is serving about 300 queries a second from interactive web users. Third-
party programmatic use of Eutils results in an additional 16 million requests
a day, producing another 450 gigabytes of data a day. Eutils use adds about
another 330 requests a second, meaning the whole Entrez framework is serving
about 600-650 requests a second.

Entrez was first released in 1991 on CD-ROM with 3 linked databases, MED-
LINE (which later became PubMed), DNA sequences (mostly GenBank) and
proteins (GenBank, Swiss-Prot, Protein Information Resource (PIR), and others). It was followed by a network client/server implementation in 1993, and the
first Web version in 1994. Entrez now provides indexing and retrieval for 40 distinct databases, including PubMed, GenBank, PubMed Central (PMC), Books,
Taxonomy, Genomes, Genes, dbSNP, and dbGaP, containing about 633 million
data records of radically different types.

O. Bodenreider and B. Rance (Eds.): DILS 2012, LNBI 7348, pp. 1–4, 2012.

2   Basic Design

Entrez was designed to federate or separate databases of different designs and implementations, containing data records with very different architectures, into a single presentation framework in which every database could be indexed and navigated in a uniform way. Entrez consists of "nodes" where each node contains records of a single uniform data type and each with a unique integer ID. The actual database containing the record normally exists outside the Entrez system, and has its own IDs such as GenBank accession numbers, that are independent of the Entrez integer IDs. An Entrez node may be populated from a single source (such as PubMed), from multiple sources (such as Gene, which gathers data from many databases, such as RefSeq and Gene Ontology Annotations), or, conversely, two Entrez nodes (such as DNA and protein) may take data from a single outside database (such as GenBank which contains both the DNA sequences of genes and the protein products of those genes).

Each Entrez node has a set of search indexes, which are essentially lists of text strings attached to Entrez unique integer IDs, and a set of DocSums (document summaries) which are short XML records keyed to the unique Entrez ID, and which contain some generic fields such as a caption and title, creation dates, etc, and unique fields in some cases such as chromosome locations or author lists.

The full record remains in the native database, and is not part of the Entrez system.

In addition, Entrez supports links between nodes, which are essentially an integer ID in one node, an integer ID in another node, a link type, and possibly a score. Through this mechanism Entrez can point between nodes (for example, the PubMed article in which a DNA sequence was published) or within the nodes of the same database (such as other PubMed articles which are similar to the PubMed article I am looking at).

The Entrez engine supports a basic set of operations on any node.

- Search with a string or a Boolean expression and return a list of integer IDs of records within a node that satisfy the query.
- Given a list of integer IDs retrieve the XML DocSums.
- Given a list of integer IDs return another list of integer IDs in another node which are linked to the input list.
- Rank and sort the lists various ways.

However, when a user wishes to view a full record from any node, Entrez passes the request back to the native database for that node. This is the point when all nodes cease to be the same, and the differences unique to each data type become important and specialized displays are necessary. However, it limits the demands on the native database, as it does not need provide high-performance or user-friendly searching or to "know about" or be linked to related resources. It also means that details of the domain specific data model for an individual database can change with the science, but the generic, high-performance retrieval system may continue unchanged.

3   Retrieval and Discovery

A data retrieval system is intended to reliably retrieve content based on the information it contains, or which is added as metadata. Considerable work has gone into Entrez to support the retrieval function with the usual methods of eliminating stop words, supporting specialized data fields, phrase matching, stemming, spell checkers, etc.

In addition, Entrez was also intentionally designed to support "discovery", or the ability to exploit relationships among data items or between databases, to add information to the system which did not exist in the original data item as deposited in the database.

The simplest and most obvious of these was establishing the link between a DNA sequence deposited into GenBank and the PubMed record for the article in which it was published. At the time Entrez was created, this link did not exist. The article citation may not have been known at the time the article was deposited into GenBank. Since GenBank focused on accuracy of DNA sequences, not accuracy or validation of article citations, many of the citations in GenBank were ill-formed or invalid, creating other linking problems. Considerable effort went into developing ways to accurately retrofit these important links as part of improving GenBank. Although the work was mostly done on the GenBank side, once the link was established from GenBank to PubMed, it is natural in the Entrez system to be able to reverse the link, and now PubMed also points to GenBank, although no changes were made to the PubMed record. As a result, PubMed users can now discover GenBank records.

A more subtle form of discovery is the concept of "neighbors", which are data items in the same database, which may have no explicit pointers between records, but which may be related to each other based on computational analysis. One of the most widely used of these neighboring links is "Related Articles" in PubMed, which are related in the sense that statistical text retrieval calculations suggest sufficient similarities in word usage in the text of the title and abstract, that the articles are likely to be discussing a similar subject. This computation is done automatically, without human indexing or curation, and in many cases brings together articles that the original authors didnt realize were related. In some cases the author couldnt know about the other article, because it was published later, but neighboring allows some viewing an older article to point "forward in time" to more recent articles on the same subject. Of course there are other reasons as well, not the least of which is the size of the biomedical literature.

Finally, Entrez supports discovery by allowing navigation between and among nodes. For example, a user may start with a PubMed record describing the function a particular protein. From there, she may link to the protein record itself, and ask to find other proteins related by sequence (protein neighbors) but not necessarily by function. From those she may return by links to PubMed to read about the functions of these related proteins and see if they may have additional functions not described in the original paper. By moving across databases like this, one can take advantage of the computed relationships unique and appropriate to one type of data, to gather and organize records of a different type of data, and, in the process, discover groups and relationships that may not be obvious or explicit in any one database.

4 Conclusion

Entrez is a search and retrieval framework which supports a large number of disparate databases for a very large number of users under production loads. It provides a uniform search and retrieval structure, while allowing the underlying databases to be uniquely structured as appropriate to that data type or activity. In addition to search and retrieval, Entrez supports computed relationships among data of a type, as well as reciprocal interlinking among data types. The power of exploration in the overall system becomes greater than the sum of its parts in this way.

Acknowledgement. This work was supported by the Intramural Research Program of NIH, NLM, NCBI.

InterOnto – Ranking Inter-Ontology Links

Silke Trißl[1], Philipp Hussels[2], and Ulf Leser[2]

[1] Leibniz Institute for Farm Animal Biology,
Wilhelm-Stahl-Allee 2, 18196 Dummerstorf, Germany
`trissl@fbn-dummerstorf.de`
[2] Humboldt-Universität zu Berlin, Institute for Computer Science,
Unter den Linden 6, 10099 Berlin, Germany
`{hussels,leser}@informatik.hu-berlin.de`

Abstract. Entries in biomolecular databases are often annotated with concepts from different ontologies and thereby establish links between pairs of concepts. Such links may reveal meaningful relationships between linked concepts, however they could as well relate concepts by chance.

In this work we present InterOnto, a methodology that allows us to rank concept pairs to identify the most meaningful associations. The novelty of our approach compared to previous works is that we take the entire structure of the involved ontologies into account. This way, our method even finds links that are not present in the annotated data, but may be inferred through subsumed concept pairs.

We have evaluated our methodology both quantitatively and qualitatively. Using real-life data from TAIR we show that our proposed scoring function is able to identify the most representative concept pairs while preventing overgeneralization. In comparison to prior work our method generally yields rankings of equivalent or better quality.

1 Introduction

Entities in many biological databases are frequently annotated with ontology concepts to assist researchers in searching, comparing, and browsing the data. Consider for instance the Arabidopsis Information Resource (TAIR) [20]. In TAIR genes are annotated with concepts from the Gene Ontology (GO) [1] and the Plant Ontology (PO) [7]. These annotations are complementary, as GO is a structured vocabulary for the functional description of genes and gene products, while PO concepts are used to describe plant structures and developmental stages.

A side effect of annotating database entries with ontology concepts is that data curators implicitly create links between concepts from different ontologies. Figure 1 shows such implicit links created by GO and PO annotations for the TAIR entry AT1G15550GA4 (gibberellin 3 β-hydroxylase). Considering these links between concepts of GO and PO a scientist may for example infer that a certain biological process is located in a specific tissue of a plant, or is active at a certain developmental stage. Take the TAIR entry for gibberellin 3 β-hydroxylase in Figure 1. Gibberellins are a family of phytohormones involved in various

O. Bodenreider and B. Rance (Eds.): DILS 2012, LNBI 7348, pp. 5–20, 2012.

developmental processes such as germination, flowering, and stem elongation in *Arabidopsis thaliana* and other vascular plants. Gibberellin 3 β-hydroxylase is an enzyme that catalyses the final biosynthetic reaction to produce bioactive gibberellin 3, an essential step in the signal cascade stimulating germination after exposure to red light [22]. This semantic relationship *red light stimulates germination* could be inferred automatically in Figure 1 by considering the link between the concepts *'response to red light'* and *'germination'*. The problem with ontology links as a means to find associated concepts is that many links are false positives. Consider again Figure 1. This TAIR entry also links the GO concepts *transcription factor binding* and *cytoplasm* to the PO concepts *root, leaf*, and *stem*. None of the six concept pairs derivable from these links is a meaningful association.

Without further preprocessing a researcher would have to look at a vast amount of ontology links to find those that are meaningful. Our sample TAIR entry is annotated with eight terms from GO and 19 terms from PO, creating 152 ontology links. In their daily work, researchers often deal with sets of dozens of genes or other biomolecular entities such as gene families, co-clustered genes in microarray experiments, or interacting proteins in PPI networks. In such a set each gene or protein may contribute unique annotations to the overall set of annotations. The resulting amount of ontology links becomes cumbersome to explore. In this work, we therefore present InterOnto, a method to rank and thus identify meaningful ontology links established by a set of database entries in an automated manner.

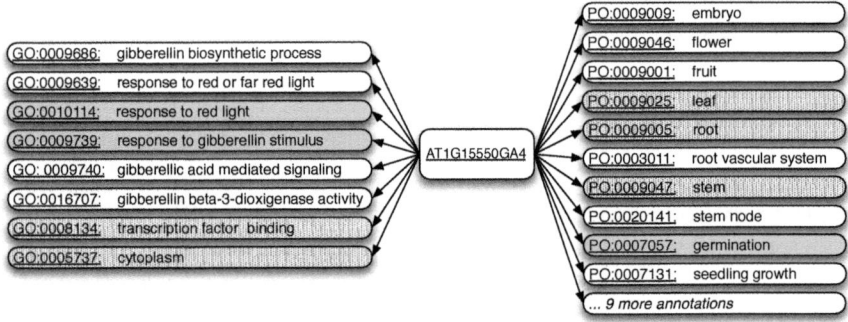

Fig. 1. Sample TAIR entry linking concepts in the Gene and Plant Ontology

The first idea to rank concept pairs that comes to mind is to count links connecting the same concepts. Certainly, the more database entries link the same concepts, the more confident we are that these links represent a meaningful relationship. The problem with this approach is that data curators describe database entries as precisely as possible, using very specific concepts. Database entries may therefore be similar and yet not share a single annotation. In these cases, the simple counting approach fails. Consider the situation depicted in Figure 2a. The three selected entries are annotated with terms from two different ontologies O_1 and O_2. These annotations establish 12 ontology links connecting 8 distinct pairs of concepts. The pairs *(e,i)*, *(g,i)*, *(g,j)* and *(f,i)* are linked twice, while the other pairs are linked once.

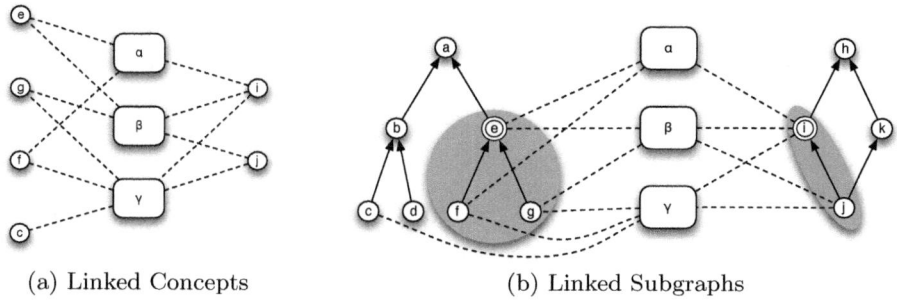

(a) Linked Concepts   (b) Linked Subgraphs

Fig. 2. Set of database entries linking concepts from two ontologies O_1 and O_2

However, the picture changes once we take not only the frequency of links, but also the relationships between ontology concepts into account. Figure 2b shows the same data plus ancestor concepts in the respective ontologies. It is evident that almost all links connect the same subgraphs in O_1 and O_2. The goal of our work is to find such pairs of strongly interlinked subgraphs and represent them with a single concept pair, i.e., the root concepts of the subgraphs. Identifying these root concepts is a challenging task. Just counting the number of links originating in a subgraph may result in overgeneralization. In Figure 2b the subgraph induced by the more general concept a contains two additional links compared to the subgraph induced by e. As result, contrary to our intuition, a would be selected as representative concept. Intuitively, the method to identify root concepts of strongly interlinked subgraphs should find a balance between the number of links in a subgraph and the generality of root concepts. One option would be to count the number of edges between the concept and the root of the ontology, i.e., the depth of a concept. This measure would assume that each edge in the ontology has the same semantic distance. As this is usually not the case, we propose to use the information content of a concept for our newly developed method InterOnto.

The remainder of this work is composed as follows. In Section 2, we present related work on identifying concept pairs from ontologies. In Section 3 we establish the basics required for our method InterOnto, which we present in Section 4. In Section 5 we present and discuss the results produced by InterOnto. Finally, in Section 6 we conclude the paper.

2   Related Work

In the first step of InterOnto we identify ontology concepts that represent a set of biological entries best. This is related to identifying significant GO concepts for a set of genes. For a review on that topic see Huang *et al.* [5]. In a different application field Brauer *et al.* [3] assign ontology concepts to text documents for ontology-supported document retrieval. To identify the most relevant concepts for a document they also consider the ontology structure and propagate scores from successor concepts upwards, which significantly improves the performance.

Extensive work has been done on finding mappings between concepts in different ontologies. Two surveys [4,9] present an overview of ontology mapping. Several approaches rely merely on concept properties, such as concept names, synonyms, or parent and child relationships [14,16]. Other approaches use instance-based ontology mapping to identify semantically equivalent concepts in different ontologies. [2,15] use association rule mining to find pairs of related GO terms given a set of entries in a database annotated with GO terms. Both approaches ignore the structure of the ontologies.

Several paper present functions to compute a similarity score for a concept pair (o_1, o_2) based on the number of instances with which o_1, or o_2, or both together are annotated with. In [10] Kirsten *et al.* present four different functions. They propagate scores to parent and grandparent concepts. In contrast to our method this approach does not consider the information loss caused through subsumption and arbitrarily limits the propagation of ontology links to only two levels. Isaac *et al.* [6] present and experimentally compare five different functions. Tan *et al.* [21] present another function to find mappings between ontology concepts based on co-occurrence in text documents. The functions in both studies may be extended to also account for ontology structure by adding the number of instances with which a descendant concept of o_1 or o_2 is annotated with.

The most similar work to our approach is LSLink [13]. LSLink uses the measures *support* and *confidence* to rank concept pairs based on links induced by selected database entries. Intuitively, the *support* for a concept pair (o_1, o_2) is a measure that gives an estimate if the number of links between o_1 and o_2 in the selected subset is statistically significant with respect to the underlying data. The *confidence* provides a measure to estimate if the number of links between o_1 and o_2 in the selected subset occurs by chance with respect to the annotations frequencies of both individual terms in the subset. Both scores are highest for a link whose concepts are annotated just once to the underlying data. In [12] the authors extend their approach from [13] by boosting the score of a parent concept by the scores of their child concepts to improve the ranking. This extended version of LSLink is similar to our approach InterOnto, but we do not restrict the score propagation to only one level.

In [18] Saha *et al.* mine the tripartite graph induced by selected database entries and their annotated ontology concepts for the densest subgraphs. To compute these subgraphs they not only consider the links themselves, but also the distance, i.e., the number of edges between concepts in an ontology, up to a certain threshold. This approach has two drawbacks. First, the edge count approach assumes that each edge represents the same semantic distance, which may not be true in many ontologies. Thus, we use the information content of a concept. Second, all relationships in a mined densest subgraph are equally important, which may still leave a researcher with a huge amount of links to explore. In contrast, we present a ranked list of links.

3   Basic Concepts

Our scoring function for mining meaningful associations from ontology links incorporates information about annotation frequencies as well as ontology structure. In this section, we briefly introduce the data and ontology model. In addition, we present a measure of semantic similarity for ontology concepts. Finally, we formalize the notion of ontology links.

3.1   Data Model and Ontology Structure

We view a data source as a comprehensive set of database entries that describe the same kind of biological entity, e.g., genes, proteins, or diseases. We assume that a particular subset of these entries, the *user dataset*, is of interest to a researcher. A user data set may for example consist of genes that were over-expressed in a microarray experiment, regulated by the same transcription factor, or associated to the same disease. Entries may be *annotated* with concepts from an *ontology*.

An ontology $O(V, E)$ is a directed acyclic graph with vertex set $V(O)$ and edge set $E(O)$. Each vertex represents a concept and edges represent *is-a* relationships between concepts. Given a concept c we use the notation $desc(c)$ and $anc(c)$ to denote the sets of descendant and ancestor concepts of c. The notations $desc'(c)$ and $anc'(c)$ are short forms for $\{c\} \cup desc(c)$ and $\{c\} \cup anc(c)$.

Given a data source S and an ontology $O(V, E)$. An annotation is an ordered pair $(s, c) \in S \times V(O)$ of a database entry and an ontology concept.

The goal of our work is to mine meaningful concept associations from sets of ontology links. In the simplest case an ontology link is established by a single entry s that is annotated with concepts from two different ontologies. Let us assume $c_1 \in V(O_1)$ and $c_2 \in V(O_2)$ are two concepts from ontologies O_1 and O_2, respectively. If annotations (s, c_1) and (s, c_2) exist, we say s establishes link (c_1, s, c_2) between the concepts c_1 and c_2.

Definition 1 (Ontology Link). *Let S be a data source and $O_1(V, E)$ and $O_2(V, E)$ be two ontologies. An ontology link $(c_1, s, c_2) \in V(O_1) \times S \times V(O_2)$ is a 3-tuple, with (c_1, s) and (c_2, s) being annotations.*

3.2   Measure of Semantic Similarity

For various practical applications it is necessary to know how similar or dissimilar two ontology concepts are [8]. Several measures have been developed to determine the semantic similarity or distance between concepts in an ontology.

The simplest distance measure between two ontology concepts is the *edge count distance* [11] where the distance is the minimum number of edges between the two concepts. The implicit assumption underlying the edge count distance is that edges in *is-a* ontologies are equidistant in terms of semantic distance. In real-world ontologies this is usually not the case. Thus, several authors [8,17] proposed to use the *self-information* of a concept to evaluate semantic similarity. The self-information is quantified as the negative logarithm of the *relative annotation frequency* $f(\hat{c})$.

Definition 2 (Self-information of a concept). *Let S be a data source, $O(V, E)$ be an ontology, and $\mathcal{A} \subset S \times V(O)$ be a set of annotations. The information content $I(\hat{c})$ of $\hat{c} \in V(O)$ regarding S is given in Equation 1.*

$$I(\hat{c}) = -\log_2(f(\hat{c})) \tag{1}$$

$f(\hat{c})$ is the relative annotation frequency of $\hat{c}$ as given in Equation 2.

$$f(\hat{c}) = \frac{|\{s \in S : \exists (s,c) \in \mathcal{A} \wedge c \in desc'(\hat{c})\}|}{|S|} \tag{2}$$

The relative annotation frequency $f(\hat{c})$ of a concept $\hat{c}$ as given in Equation 2 is the relative frequency of entries with which the concept itself or one of its sub-concepts c is annotated.

Resnik [17] suggested to use the information content of the most informative common ancestor of two ontology concepts c_1 and c_2 as measure of semantic similarity of two concepts. This score is called *shared information content.*

Definition 3 (Shared information content). *Let $c_1, c_2 \in V(O)$ be two concepts in an ontology O. Let $C = anc'(c_1) \cap anc'(c_2)$ denote the set of common ancestor concepts of c_1 and c_2. The shared information content σ of c_1 and c_2 is:*

$$\sigma(c_1, c_2) = \max_{\hat{c} \in C} I(\hat{c}) \tag{3}$$

Note, the measure σ does not produce values between 0 and 1. The scores range from 0 for the root concept to a maximum value for a given set of annotations. A concept obtains this maximum value if it is present in the least number of annotations, which usually means in exactly one annotation. This fact is important for understanding the results. An advantage of this measure is that it can naturally be extended to determine the similarity of concept sets of arbitrary size, as long as these concepts share a common ancestor. In Section 4 we show how this property is particularly useful for our application.

4   InterOnto – Linking Ontologies Using Evidence

In the following sections we present InterOnto, a methodology to rank pairs of ontology concepts based on how likely they represent meaningful information to a researcher. We show how to incorporate information encoded in the structure of *is-a* ontologies to improve the rankings.

Consider Figure 2b again. When we simply count the links we get scores for the leaf concept pairs (f, j) and (g, j) of 1 and 2, respectively. However, an entry annotated with a particular ontology concept is implicitly also annotated with all its ancestor concepts. Thus, the scores for concept pairs (f, i), (g, i), and (e, j) are 3, 4, and 4 when counting the number of links in the subgraphs. The pair (e, i) outscores these pairs with 10 supporting links. Eventually, this approach would choose the concept pair (a, i) as it has the highest score of 12. This is counterintuitive as from a user's perspective the concept pair (e, i) would represent the selected subset best. The reason for choosing (a, i) over (e, i) is, that the counting approach does not take the loss of specificity caused by subsumption into account.

4.1   Finding Representative Concepts

Consider the situation depicted on the left side of Figure 3. Intuitively, we can identify two distinct groups of two and three annotations, as highlighted in grey. To represent these groups, one would probably choose the concepts a and d as representative concepts. Counting the number of annotations of a concept and its successor concepts would rank concept a highest, as it has five annotations. The counting approach does not consider the loss of specificity when moving up the ontology. To model this loss, we propose Equation 4 to assign a *similarity based score* to a concept $\hat{c}$ with respect to the set of annotations $\mathcal{A}$ present in the user dataset.

Definition 4 (Similarity based scoring function). *Let S' be the user selected dataset of data source S, $O(V, E)$ an ontology, and $\mathcal{A}' \subset S' \times V(O)$ a set of annotations. The similarity based score of a concept $\hat{c} \in V(O)$ is given by:*

$$score_\sigma(\hat{c}) = \sum_{c \in desc'(\hat{c})} |\{(s, c) : (s, c) \in \mathcal{A}'\}| \cdot \sigma(c, \hat{c}) \tag{4}$$

If we omit factor $\sigma(c, \hat{c})$ in Equation 4 we obtain the number of annotations in the subgraph of $\hat{c}$. The factor $\sigma(c, \hat{c})$ describes the similarity between concepts c and $\hat{c}$, i.e., the more similar the concepts are the higher the similarity value.

In Equation 4 the contribution of a specific annotation c to the overall score of a concept $\hat{c}$ is the shared information content of c and $\hat{c}$, $\sigma(c, \hat{c})$. According to Definition 3 this score equals $\max_{\hat{c} \in C} I(\hat{c})$. Since $\hat{c}$ is per definition the most specific common ancestor of all c we may simply use the self-information of $\hat{c}$, $I(\hat{c})$. This allows us to transform Equation 4 to Equation 5.

$$score_\sigma(\hat{c}) = |\{(s, c) : (s, c) \in \mathcal{A}' \wedge c \in desc'(\hat{c})\}| \cdot I(\hat{c}) \tag{5}$$

In Equation 5 the score of a given concept depends on its self-information and the number of annotations in which it is present in the user dataset. Consider the example in Figure 3 again. Let us assume that each depicted concept is annotated to one entry in the underlying data source and the total number of annotations for Ontology 1 and 2 is 50 each. Thus, for concepts d and a we have an information content of $I(d) = -\log_2 \frac{3}{50} = 4.06$ and $I(a) = -\log_2 \frac{14}{50} = 1.84$. Calculating $score(\hat{c})$ for all concepts results in the highest score for concept d with $3 \cdot 4.06 = 12.2$, followed by c with $3 \cdot 3.64 = 10.9$ and a with $5 \cdot 1.84 = 9.2$. This example as well as our experimental evaluation shows that $score_\sigma$ allows us to identify representative concepts, but overcomes the problem of overgeneralization; the more general a representative concept is, the more annotations must support it to yield a good score, as per definition $I(\hat{c}) \leq I(c), \forall \hat{c} \in anc(c)$.

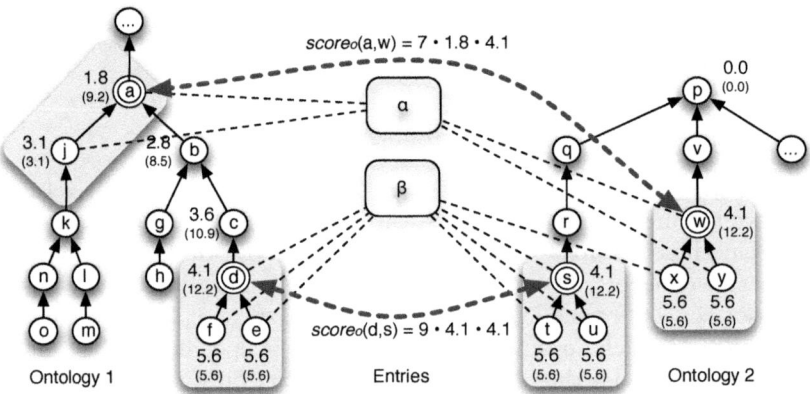

Fig. 3. Scores $score_\sigma(\hat{c}_1, \hat{c}_2)$ for concept pairs. The numbers on nodes represent the information content $I(\hat{c})$ of a concept, while the numbers in brackets represent $score_\sigma(\hat{c})$.

4.2   Scoring Representative Concept Pairs

We now get back to the original problem of mining meaningful associations from ontology links. An initial idea would be to produce the cartesian product of concepts in O_1 and O_2 and rank the concept pairs based on the product of $score_\sigma(\hat{c}_1)$ and $score_\sigma(\hat{c}_2)$. However, this approach does not yield a meaningful result as it does not consider the number of links established through entries in the user dataset. Consider the example shown in Figure 3. The three most informative concepts are d, s, and w. The cartesian product would yield the same score for concept pairs (d, s) and (d, w). However, intuitively we would rank (d, s) higher as it is supported by nine links in total (one direct link over entry β and eight through descendant concepts of d and s over β). In comparison, for concept pair (d, w) we find three supporting links, namely (d, β, x), (f, β, x), and (e, β, x).

The example shows that simply multiplying the similarity scores does not yield the desired result to find representative concept pairs. We may not just consider annotations (s, c_1) and (s, c_2) separately, but we have to consider the actual links between concepts in O_1 and O_2 established through entries in the user dataset. Equation 6 provides the similarity based scoring function for pairs of ontology concepts (c_1, c_2).

Definition 5 (Similarity based scoring function). *Let S' be the user selected dataset of data source S and O_1 and O_2 two ontologies. Let furthermore $\mathcal{L}' \subset V(O_1) \times S' \times V(O_2)$ denote a set of ontology links between them. The similarity based score of a concept pair $(\hat{c}_1, \hat{c}_2)$ with $\hat{c}_1 \in V(O_1)$ and $\hat{c}_2 \in V(O_2)$ is given by Equation 6.*

$$
\begin{aligned}
score_\sigma(\hat{c}_1, \hat{c}_2) = & |\{(c_1, s, c_2) : (c_1, s, c_2) \in \mathcal{L}' \wedge c_1 \in desc'(\hat{c}_1) \wedge c_2 \in desc'(\hat{c}_2)\}| \\
& \cdot I(\hat{c}_1) \cdot I(\hat{c}_2)
\end{aligned}
$$

$$(6)$$

4.3 Eliminating Redundant Representative Concept Pairs

While incorporating hierarchy information may improve the result set, it may also introduce a significant number of redundant concept pairs. Consider for example the concept pair (c, s) in Figure 3. This pair represents the same set of links as the pair (d, s), but is less specific and therefore receives a lower score. When evaluating $score_\sigma(c_1, c_2)$ in Section 5 we thus eliminate pairs that do not add information to the overall set of concept pairs. For every set of links we only keep the most informative representative concept pair, i.e., the pair that receives the highest score.

5 Evaluation

We tested our methodology on several real world data sets, which we present in Section 5.1. We compare the top-k concept pairs produced by our method InterOnto with the top-k concept pairs resulting from the LSLink measures *support* and *confidence*.

5.1 Input Data

In the following subsections we give an overview of the ontologies, biological data sources, and data sets used for evaluation purposes.

Ontologies. In our experiments we used the Gene Ontology (GO) [1] and the Plant Ontology (PO)[7]. The Gene Ontology contains three different sub-ontologies, namely *'Molecular Function'*, *'Biological Process'*, and *'Cellular Component'*. The Plant Ontology contains two different sub-ontologies, which are *'Plant Structure'* and *'Plant Growth and Development Stage'*. In our experiments we consider these different sub-ontologies as independent ontologies, as no *'is-a'* relationships between concepts in these sub-ontologies exist.

Biological Data Sources. We used TAIR [20] as data source. In TAIR an entry may be annotated with concepts from GO and PO. We found that 52,766 entries in TAIR are annotated with GO concepts and 19,883 entries with PO concepts. In total we found 145,627 GO-TAIR annotations and 514,567 PO-TAIR annotations, resulting in 3,361,887 distinct GO-PO concept pairs.

Test Data Sets. For our quantitative evaluation we selected 20 gene sets from TAIR that constitute gene families, shown in Table 1. The genes in these gene families fulfill a similar role in the organism. Thus, for manual inspection we are able to evaluate if high ranked concepts and concept pairs are meaningful.

Table 1. Selected gene families from the TAIR database

id	Gene Family Name	Genes	Annotated[%]	Annotations	Unique
1	Core Cell Cycle Genes	61	98	303	102
2	basic Helix-Loop-Helix (bHLH) Transcription Factor	162	98	565	103
3	Plant Cell Wall Biosynthesis Families	31	97	167	36
4	Cytoplasmic ribosomal protein gene family	248	95	1090	48
5	Lipid Metabolism Gene Families	98	94	280	98
6	Chloroplast and Mitochondria gene families	50	94	222	53
7	Primary Pumps (ATPases) Gene Families	81	89	253	104
8	Monosaccharide transporter-like gene family	53	100	222	40
9	Acyl Lipid Metabolism Family	610	92	1802	425
10	Kinesins	61	98	122	40
11	zinc finger-homeobox gene family	17	94	45	14
12	Glycosyltransferase Gene Families	280	98	1032	171
13	ABC Superfamily	126	97	307	96
14	Heat Shock Transcription Factors	21	100	77	20
15	Protein synthesis factors	95	99	321	64
16	Inorganic Solute Cotransporters	83	100	323	86
17	Ion Channel Families	59	100	240	58
18	Phosphoribosyltransferases (PRT)	15	100	58	23
19	Glycoside Hydrolase Gene Families	307	98	664	158
20	Response Regulator	32	100	197	38

5.2   Evaluation Method

Relating concepts from orthogonal ontologies is a relatively new research area. To our knowledge no gold standard exists, with which we could compare our results. One option to assess the quality of our ranking is by domain experts. But this assessment is very time consuming and may be subjective. We thus decided to use existing inter-ontology mappings as basis for our evaluation.

The problem with existing inter-ontology mappings is that they are usually not very comprehensive. For instance, the mapping between PO and GO provided by the OBO Foundry [19] consists of only 137 relationships between the sub-ontologies *Biological Process* and *Plant Structure*. A notable exception to this lack of coverage are the inter-ontology mappings in GO itself, where the sub-ontologies *Molecular Function* and *Biological Process* are richly inter-linked. In our snapshot of GO we found 517 relationships of type *biological process regulates molecular function* and 206 relationships of type *molecular function is part of biological process*. At first glance, this number seems sufficient for assessing the ability of methods to rediscover those links.

Despite these numbers we still face the problem that many established mappings are fairly generic. Consider for example the term pair *('transferase activity, transferring glycosyl groups', 'polysaccharide biosynthetic process')*. In fact glycosyl transferases are key enzymes in the synthesis of polysaccharides, but no valid path in GO confirms this fact. The closest established relationship, as depicted in Figure 4, is the fairly generic association *'catalytic activity' part-of 'metabolic process'*.

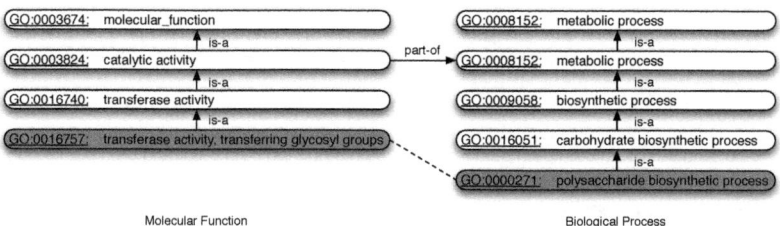

Fig. 4. Example for a potentially true positive association, PTP (dotted line)

To also use such generic relationships for assessing more specific concept pairs, we introduce *Potentially True Positive associations* (PTPs). A potentially true positive association between two ontology concepts exists, if the concepts themselves or any of their ancestor concepts have an established inter-ontology mapping.

5.3   Ranking Representative Concept Pairs

We now evaluate the quality of proposed inter-ontology concept pairs. We validate the top-k concept pairs found by our method InterOnto and by LSLink against existing inter-ontology mappings. This allows us to compare both methods quantitatively.

We used the 20 gene families from TAIR to compute sets of inter-ontology links using our own scoring function $score_\sigma$ and the *confidence* and *support* scores defined by the LSLink methodology [13]. As baseline we used randomly ordered ontology links of each set. To compare the different result sets we partitioned each ranking into 10 equally sized sublists and extracted the first 10 entries from each sublist, starting with the top-10 concept pairs. Figure 5 shows the average number of PTPs for each of the 10 sublists averaged over all 20 gene families.

The most notable finding is that on average the top-10 lists for all three scoring functions contain considerably more PTPs than subsequent or random samples. This effect is stronger for $score_\sigma$ and *confidence* than for the *support* score. Since the PTP heuristic is based on well-known relationships the different results for *support* and *confidence* are actually in accordance with our expectations. The overall correlation of rank and number of PTPs is clearly the strongest for $score_\sigma$.

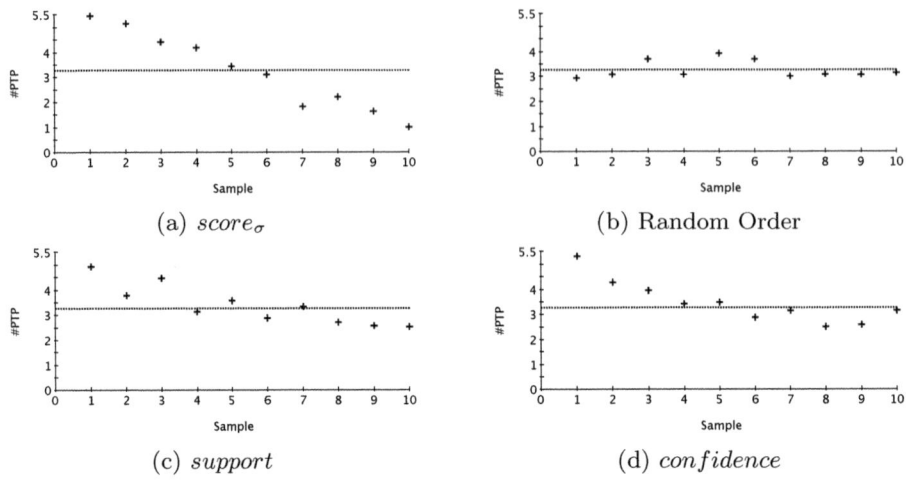

Fig. 5. Number of PTPs in top-10 concept pairs from 10 equally sized partitions. Samples are numbered in ascending order of partition by rank. Values are averaged over 20 different sets of links between GO Molecular Function and GO Biological Process, corresponding to the 20 TAIR gene families. Dotted lines mark the average number of PTPs among random concept pairs.

Using linear regression we yield a fit line with an incline of -0.5 and a regression factor of 0.97. For *confidence* and support we measure an incline of -0.242 and -0.236 and regression factors of 0.77 and 0.74, respectively.

The plots in Figure 5 do in principle confirm our expectations, although we would have expected a higher drop between the top-10 links and subsequent or random samples. One reason for not observing that drop is the presence of overly general mappings in the Gene Ontology, e.g., 'enzyme regulator activity' regulates 'catalytic activity'. These mappings may detect meaningless PTPs and thus increase the score for subsequent or random samples. Another factor influencing the results is that some meaningful relationships are not modeled at all. Considering the results for the different gene families from TAIR depicted in Figure 6 confirms our assumption. The plots show a strong variation in number of PTPs, with some top-10 lists not containing a single PTP for all three scores. Take for example the gene family '4 - Kinesins'. Kinesins are motor proteins that move along microtubules. Our top-10 list contains meaningful term pairs such as ('microtubule motor activity', 'microtubule based movement'). Yet, none of these pairs is a PTP.

The results of our analysis suggest that our approach generally yields rankings of higher quality than those produced by LSLink. We further studied how the actual top-k concept pairs differ. We determined the relative overlap of top-5 and top-10 lists for all three functions. We found that the lists produced by LSLink's *confidence* and *support* scores overlap to a much higher degree with each other than with results produced using $score_\sigma$ (data not shown).

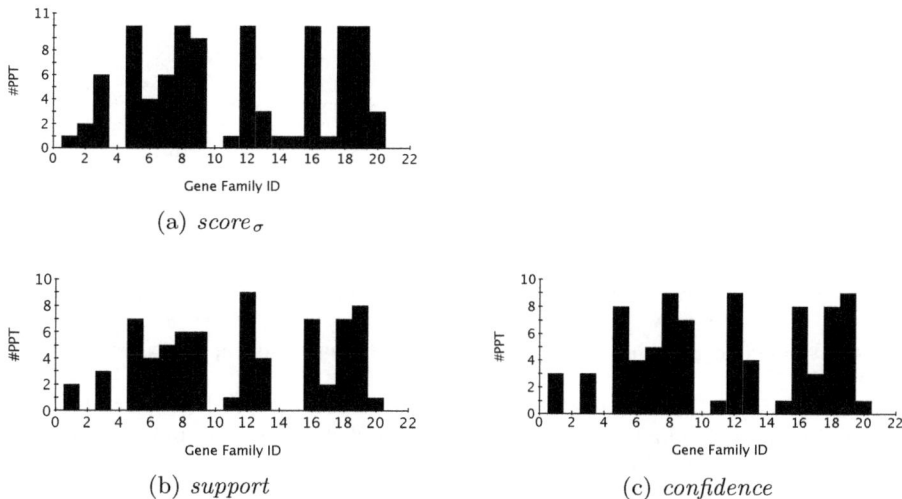

Fig. 6. Number of potential true positive (PTP) associations in top-10 concept pairs for different link sets, generated using different scoring functions

Manual analysis of sample results showed that we may attribute these differences to the impact of hierarchy information. To quantify the differences we determined the information content of associated concepts and visualized them in Figure 7 in separate 2-dimensional plots for the top-k concept pairs computed by each of the three scoring functions (plot for *confidence* similar to plot for *support* and thus omitted). For easier interpretation we added dashed lines to depict the maximum values for self-information. Note that, a concept obtains the maximum value only if it is annotated in this scenario to exactly one database entry. The plots for the top-5 concept pairs produced by the measures *confidence* and *support* (Figure 7b) show that many of the top-ranked pairs contain at least one concept with maximum self-information. This is counter-intuitive, since such associations are based only on a single ontology link, which may be a coincidence or an error in annotated data. In contrast, Figure 7a shows that our approach does not yield such low evidence concept pairs. We may assume though that if such associations are indeed meaningful, several similar links exist. In this case, our method should return a more general concept pair with lower self-information score that subsumes these highly specific links. Our analysis shows that in InterOnto on average 54% of the top-5 and 62% of top-10 concept pairs are inferred through subsumption and have thus not been present in the original links or can be detected without hierarchy-aware methods. This way, InterOnto may be more robust to errors in annotations compared to LSLink. Another desirable property of InterOnto is that it is more likely to associate concepts of similar specificity, as the majority of points are distributed along the graphs main diagonal.

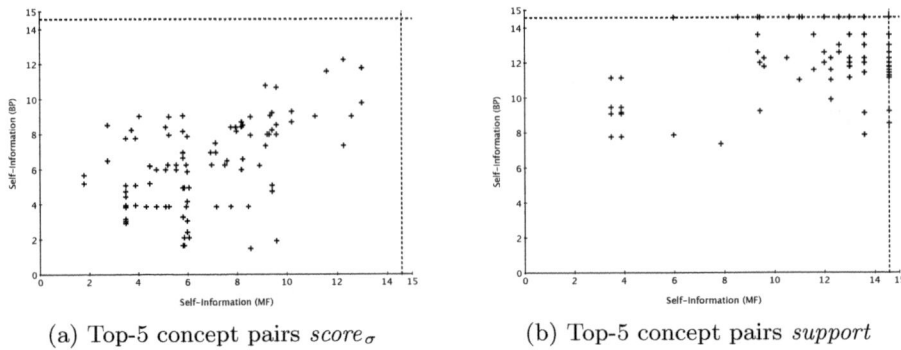

(a) Top-5 concept pairs *score_σ*   (b) Top-5 concept pairs *support*

Fig. 7. Self-Information of paired GO and PO concepts in top-5 lists generated using different measures. Evaluation was performed on 20 distinct sets of links corresponding to the 20 TAIR gene families listed in Table 1.

Table 2. Top-5 concept pairs for TAIR gene family 16 - 'Inorganic Solute Cotransporters'

#	Molecular Function	Biological Process
	Top-5 pairs for *score_σ*	
1	ion transmembrane transporter activity	ion transport
2	substrate-specific transmembrane transporter activity	ion transport
3	transmembrane transporter activity	ion transport
4	ion transmembrane transporter activity	cation transport
5	metal ion transmembrane transporter activity	cation transport
	Top-5 pairs for *support*	
1	molybdate ion transmembrane transporter activity	molybdate ion transport
2	high affinity copper ion transmembrane transporter activity	high-affinity copper ion transport
3	high affinity secondary active ammonium transmembrane transporter activity	ammonium transport
4	ammonium transmembrane transporter activity	ammonium transport
5	ammonium transmembrane transporter activity	methylammonium transport
	Top-5 pairs for *confidence*	
1	molybdate ion transmembrane transporter activity	molybdate ion transport
2	high affinity copper ion transmembrane transporter activity	high-affinity copper ion transport
3	high affinity secondary active ammonium transmembrane transporter activity	ammonium transport
4	low affinity phosphate transmembrane transporter activity	phosphate transport
5	ammonium transmembrane transporter activity	ammonium transport

As a concrete example we use the TAIR gene family 16 'Inorganic Solute Cotransporters'. Table 2 shows the top-5 concept pairs for all three functions. Notably, all top-5 lists contain only biologically valid associations. The crucial difference is that the pairs ranked high by LSLink methods all refer to the transport of specific substances through the cell membrane, while our approach returns pairs that refer to ion transport through the cell membrane in general. In other words, LSLink returns specific examples of the information contained in the link sets, while InterOnto summarizes this information, characterizing the overall dataset based on evidence provided by a large number of similar links.

6   Conclusion

We introduced a new scoring function to rank concept associations from a set of ontology links. In contrast to existing work our approach considers not only ontology concepts linked directly, but also the hierarchy of ontologies in a systematic manner. Our results show that incorporating hierarchy information allows the identification of more descriptive, yet not over-general, concept pairs compared to methods that do not incorporate hierarchy information, such as LSLink. Based on our experiments we believe that our method performs well and should be useful for researchers. For a thorough evaluation of our and other methods linking ontology concepts a gold standard for the quality of concept pairs would be useful.

Acknowledgement. This work was partly supported by the BMBF supported project Phänomics (grant no. 035539G). We would like to thank Louiqa Rashid for fruitful discussions on the topic of ontology links.

References

1. Ashburner, M., Ball, C., Blake, J., Botstein, D., et al.: Gene ontology: tool for the unification of biology. The Gene Ontology Consortium. Nature Genetics 25(1), 25–29 (2000)
2. Bodenreider, O., Aubry, M., Burgun, A.: Non-lexical approaches to identifying associative relations in the gene ontology. In: Pac. Symp. Biocomput., pp. 91–102 (2005)
3. Brauer, F., Huber, M., Hackenbroich, G., Leser, U., et al.: Graph-Based Concept Identification and Disambiguation for Enterprise Search. In: Proceedings of the 19th International Conference on World Wide Web (WWW), pp. 171–180. ACM (2010)
4. Castano, S., Ferrara, A., Montanelli, S., Varese, G.: Ontology and Instance Matching. In: Paliouras, G., Spyropoulos, C.D., Tsatsaronis, G. (eds.) Multimedia Information Extraction. LNCS, vol. 6050, pp. 167–195. Springer, Heidelberg (2011)
5. Huang, D.W., Sherman, B.T., Lempicki, R.A.: Bioinformatics enrichment tools: paths toward the comprehensive functional analysis of large gene lists. Nucleic Acids Res. 37(1), 1–13 (2009)
6. Isaac, A., van der Meij, L., Schlobach, S., Wang, S.: An Empirical Study of Instance-Based Ontology Matching. In: Aberer, K., Choi, K.-S., Noy, N., Allemang, D., Lee, K.-I., Nixon, L.J.B., Golbeck, J., Mika, P., Maynard, D., Mizoguchi, R., Schreiber, G., Cudré-Mauroux, P. (eds.) ASWC 2007 and ISWC 2007. LNCS, vol. 4825, pp. 253–266. Springer, Heidelberg (2007)
7. Jaiswal, P., Avraham, S., Ilic, K., Kellogg, E.A., et al.: Plant Ontology (PO): a Controlled Vocabulary of Plant Structures and Growth Stages. Comp. Funct. Genomics 6(7-8), 388–397 (2005)
8. Jiang, J.J., Conrath, D.W.: Semantic Similarity Based on Corpus Statistics and Lexical Taxonomy. In: Proceedings of the Tenth International Conference on Research on Computational Linguistics (ROCLING), pp. 19–33 (1997)
9. Kalfoglou, Y., Schorlemmer, M.: Ontology mapping: the state of the art. The Knowledge Engineering Review 18(1), 1–31 (2003)

10. Kirsten, T., Thor, A., Rahm, E.: Instance-Based Matching of Large Life Science Ontologies. In: Cohen-Boulakia, S., Tannen, V. (eds.) DILS 2007. LNCS (LNBI), vol. 4544, pp. 172–187. Springer, Heidelberg (2007)
11. Lee, J.H., Kim, M.-H., Lee, Y.-J.: Ranking Documents in Thesaurus-Based Boolean Retrieval Systems. Inf. Process. Manage. 30(1), 79–92 (1994)
12. Lee, W.-J., Raschid, L., Sayyadi, H., Srinivasan, P.: Exploiting Ontology Structure and Patterns of Annotation to Mine Significant Associations between Pairs of Controlled Vocabulary Terms. In: Bairoch, A., Cohen-Boulakia, S., Froidevaux, C. (eds.) DILS 2008. LNCS (LNBI), vol. 5109, pp. 44–60. Springer, Heidelberg (2008)
13. Lee, W.-J., Raschid, L., Srinivasan, P., Shah, N., Rubin, D., Noy, N.: Using Annotations from Controlled Vocabularies to Find Meaningful Associations. In: Cohen-Boulakia, S., Tannen, V. (eds.) DILS 2007. LNCS (LNBI), vol. 4544, pp. 247–263. Springer, Heidelberg (2007)
14. Maedche, A., Staab, S.: Measuring Similarity between Ontologies. In: Gómez-Pérez, A., Benjamins, V.R. (eds.) EKAW 2002. LNCS (LNAI), vol. 2473, pp. 251–263. Springer, Heidelberg (2002)
15. Myhre, S., Tveit, H., Mollestad, T., Laegreid, A.: Additional Gene Ontology structure for improved biological reasoning. Bioinformatics 22(16), 2020–2027 (2006)
16. Noy, N., Musen, M.: The PROMPT suite: Interactive tools for ontology merging and mapping. International Journal of Human-Computer Studies 59(6), 983–1024 (2003)
17. Resnik, P.: Using Information Content to Evaluate Semantic Similarity in a Taxonomy. In: Proceedings of the 14th International Joint Conference on Artificial Intelligence (IJCAI), pp. 448–453 (1995)
18. Saha, B., Hoch, A., Khuller, S., Raschid, L., Zhang, X.-N.: Dense Subgraphs with Restrictions and Applications to Gene Annotation Graphs. In: Berger, B. (ed.) RECOMB 2010. LNCS, vol. 6044, pp. 456–472. Springer, Heidelberg (2010)
19. Smith, B., Ashburner, M., Rosse, C., Bard, J., et al.: The OBO Foundry: coordinated evolution of ontologies to support biomedical data integration. Nat. Biotechnol. 25(11), 1251–1255 (2007)
20. Swarbreck, D., Wilks, C., Lamesch, P., Berardini, T.Z., et al.: The Arabidopsis Information Resource (TAIR): gene structure and function annotation. Nucleic Acids Research 36(Database issue), D1009–D1014 (2008)
21. Tan, H., Jakonienė, V., Lambrix, P., Aberg, J., Shahmehri, N.: Alignment of Biomedical Ontologies Using Life Science Literature. In: Bremer, E.G., Hakenberg, J., Han, E.-H(S.), Berrar, D., Dubitzky, W. (eds.) KDLL 2006. LNCS (LNBI), vol. 3886, pp. 1–17. Springer, Heidelberg (2006)
22. Yamaguchi, S., Smith, M.W., Brown, R.G., Kamiya, Y., Sun, T.: Phytochrome Regulation and Differential Expression of Gibberellin 3β-Hydroxylase Genes in Germinating Arabidopsis Seeds. Plant Cell 10(12), 2115–2126 (1998)

Finding Cross Genome Patterns
in Annotation Graphs

Joseph Benik[1], Caren Chang[1], Louiqa Raschid[1], Maria-Esther Vidal[2],
Guillermo Palma[2], and Andreas Thor[3]

[1] University of Maryland, USA
[2] Universidad Simón Bolívar, Venezuela
[3] University of Leipzig, Germany
{jvb,louiqa}@umiacs.umd.edu, carenc@umd.edu, {mvidal,gpalma}@ldc.usb.ve,
thor@informatik.uni-leipzig.de

Abstract. Annotation graph datasets are a natural representation of scientific knowledge. They are common in the life sciences where concepts such as genes and proteins are annotated with controlled vocabulary terms from ontologies. Scientists are interested in analyzing or mining these annotations, in synergy with the literature, to discover patterns. Further, annotated datasets provide an avenue for scientists to explore shared annotations across genomes to support cross genome discovery. We present a tool, PAnG (Patterns in Annotation Graphs), that is based on a complementary methodology of graph summarization and dense subgraphs. The elements of a graph summary correspond to a pattern and its visualization can provide an explanation of the underlying knowledge. We present and analyze two distance metrics to identify related concepts in ontologies. We present preliminary results using groups of Arabidopsis and C. elegans genes to illustrate the potential benefits of cross genome pattern discovery.

1 Introduction

Arabidopsis thaliana is a flowering plant that is widely used as a model organism and whose genome was completely sequenced in the year 2000. The Arabidopsis Information Resource (TAIR) is a well curated and heavily used portal for accessing Arabidopsis genome information [6,19,21]. TAIR provides a rich synopsis of each gene through links to a variety of data including Gene Ontology (GO) [2,7] and Plant Ontology (PO) [26].

We illustrate annotation datasets using a study of genes involved in photomorphogenesis. The GO-PO annotation graph for gene CRY2 is in Figure 1. The PO annotations for CRY2 are on the left side and the GO annotations are on the right. We label this a *tri-partite annotation graph* or *TAG*. Each node of the *TAG* includes the identifier and the label for the Controlled Vocabulary (CV) term. As of September 2011, there were 17 GO and 37 PO annotations for CRY2. The figure illustrates partial annotations. On the right of Figure 1 is a fragment of the relevant GO ontology.

O. Bodenreider and B. Rance (Eds.): DILS 2012, LNBI 7348, pp. 21–36, 2012.
© Springer-Verlag Berlin Heidelberg 2012

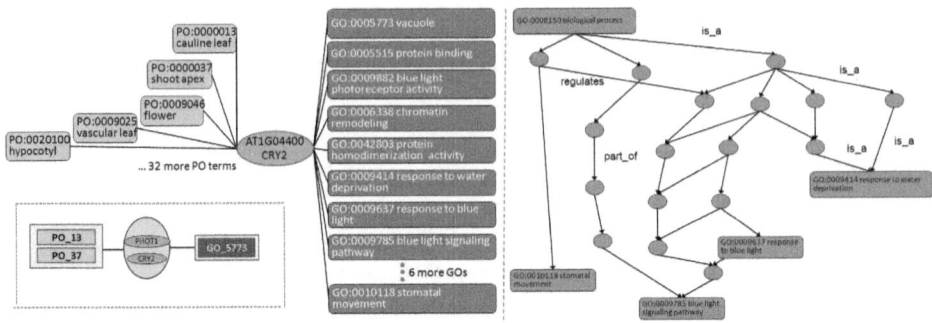

Fig. 1. GO and PO annotations for gene CRY2 (middle); GO fragment (right); Graph Summary(GS) for genes CRY2 and PHOT1 (inset)

Over the past 25 years, knowledge of the Arabidopsis genome has increased exponentially, together with that of other model organisms. This abundance of data has led to an era of comparative genomics, in which genes can be compared across diverse taxa to provide insights into evolutionary similarities as well as key divergences. Already the study of genes in Arabidopsis has helped to inform human research and vice versa. Increasingly, every new genome must be understood in light of previously sequenced and analyzed genomes.

We will consider orthologous genes from three model organisms: *Saccharomyces cerevisiae* (yeast), *Caenorhabditis elegans* (nematode) and *Drosophila melanogaster* (fruit fly). The rationale for including these highly annotated genomes is to synthesize existing knowledge to enhance our understanding of gene function in Arabidopsis (and possibly vice versa). A future aim of our research will be to extend cross-genome analysis to include a range of plants, including lower plant species. Currently, a number of these genomes are sparsely annotated. By incorporating the genomes of such plant species, we may help to build knowledge in less well-studied species by bootstrapping to Arabidopsis, while strengthening our overall understanding of plant gene function and evolution.

We recognize that our goals are ambitious and that we have to solve numerous challenges. First, we have to find patterns in annotation graph datasets. On this challenge, we can report some initial success [1,22,27].

Next, we must integrate annotation data across multiple organisms to perform comparative genomics. We must identify a protocol and efficient processes to obtain orthologs or other matching genes from multiple organisms. Potential resources and tools include the Homologene service from NCBI [9], Inparanoid [10], a database that includes animals, and Plaza [18], which is exclusive to plant species. In this paper, we bypass this potentially expensive process and describe a simpler, less expensive protocol. We use shared annotations, and gene and protein families, to harvest Arabidopsis and *C. elegans* genes and annotations for cross-genome analysis. We recognize that this protocol is less accurate at finding orthologous genes and we use it only for proof-of-concept purposes.

A key element in finding patterns is identifying related ontological concepts. A fragment of the GO ontology is shown on the right of Figure 1. We postulate that ontology terms that are located in proximity to each other in the ontology are more related. In addition, terms which are located along branches of the tree with greater depth and/or breadth potentially reside in areas where the ontological concepts are defined at a more granular level of functional or descriptive detail. Finally, pairs of terms within the same proximity and that are (both) more distant from the root of the tree may be more related. We propose a metric d_{tax} for taxonomic distance and compare to d_{ps} [16], a state-of-the-art metric. Figure 1 also illustrates different types of relationships in GO including *part_of*, *is_a* and *regulates*. While these relationship types are important in determining relatedness, we have not used these features in our current work. The contributions of this paper are as follows:

- We present the concept of tripartite annotation graphs (TAG) and our tool PAnG (Patterns in Annotation Graphs) to identify patterns. PAnG relies on dense subgraphs and graph summarization methods.
- Using sample datasets of groups of genes from Arabidopsis and *C. elegans* that share similar gene function, we show some preliminary results of validating PAnG for cross genome analysis.
- We study the properties of metrics d_{tax} and d_{ps} for a subset of GO terms and demonstrate that d_{tax} is better able to discriminate between taxonomically close terms.

This paper is organized as follows: Section 2 presents an overview of PAnG including dense subgraphs and graph summarization. Section 3 considers groups of genes from Arabidopsis and *C. elegans*, with shared function and GO annotation, to explore the potential benefits of cross-genome pattern discovery. Section 4 presents the two distance (similarity) metrics d_{tax} and d_{ps} and compares their properties on several subsets of GO biological process (GO-BP) terms.

2   Overview of PAnG

Figure 2 illustrates the overall workflow of PAnG. The input is a tripartite annotated graph G, and the output is a graph summary. Our workflow consists of two steps. The first step is optional and deals with the identification of dense subgraphs, i.e., highly connected subgraphs of G that are (almost) cliques. The goal is to identify interesting regions of the graph by extracting a relevant subgraph.

Next, graph summarization transforms the graph into an equivalent compact graph representation. Graph summaries are made up of the following elements: (1) supernodes; (2) superedges; (3) deletion and addition edges (corrections). The left inset of Figure 1 shows a fragment of a graph summary obtained from the analysis of photomorphogeneis genes in Arabidopsis. There is a supernode with the two genes PHOT1 and CRY2 and another supernode with two PO terms. There is a superedge between these two supernodes reflecting that the two genes are both annotated with the two PO terms. Both genes are also annotated with

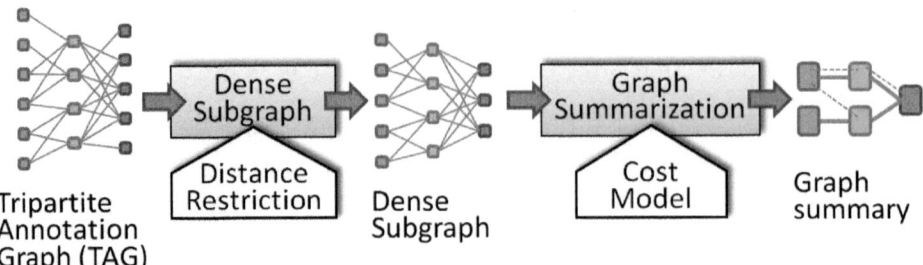

Fig. 2. The original TAG can be subject to an optional filter step to identify dense subgraphs. The PAnG tool employs graph summarization to identify patterns.

the GO term. We note that while this appears to be a very simple pattern, the association of these two genes and their PO and GO terms annotations represented an as yet unknown and potential interaction between phototropins (PHOT1) and chryptochromes (CRY2) [22].

The summary reflects the basic pattern (structure) of the graph and is accompanied by a list of corrections, i.e., deletions and additions, that express differences between the graph and its simplified pattern. For example, a deletion reflects that a gene *does not have* a particular annotation that is shared by other genes within the supernode.

A graph summary has several advantages. First, it gives a better understanding of the structure of the underlying graph and is good for visualization. Second, the summary captures semantic knowledge not only about individual nodes and their connections but also about groups of related nodes. Third, the corrections, in particular deletions, are intuitive indicators for future edge prediction.

Our approach is not limited to TAGs. A k-partite layered graph can be first converted to a more general (bi-partite) graph. Our experience is that when presented with patterns, a bi-partite graph that combines terms from multiple ontologies into one layer may not convey the same intuitive meaning to a scientist. With more than 3 layers, however, the patterns become more difficult to comprehend.

2.1 Dense Subgraphs

Given an initial tripartite graph, a challenge is to find interesting regions of the graph, i.e., candidate subgraphs, that can lead to valuable patterns. We start with the premise that an area of the graph that is rich or dense with annotations is an interesting region to identify candidate subgraphs. For example, for a set of genes, if each is annotated with a set of GO terms and/or a set of PO terms, then the set of genes and GO terms, or the set of genes and PO terms, form a clique. We thus exploit cliques, or dense subgraphs (DSG) representing cliques with missing edges. Density is a measure of connectedness. It is the ratio of the number of induced edges to the number of vertices in the subgraph.

Even though there is an exponential number of subgraphs, a subgraph of maximum density can be found in polynomial time [13,8,5]. In contrast, the maximum clique problem to find the subgraph of largest size having all possible edges is NP-hard; it is even NP hard to obtain any non-trivial approximation. Finding densest subgraphs with additional size constraints is NP hard [12]; yet, they are more amenable to approximation than the maximum clique problem.

An annotation graph is a tripartite graph $G = ((A, B, C), (X, Y))$. PAnG employs our approach in [22] and thus first transforms the tripartite graph G in a weighted bipartite graph $G' = (A, C, E)$ where each edge $e = (a, c) \in E$ is labeled with the number of nodes $b \in B$ that have links to both a and c. We then compute a densest bipartite subgraph G_2 by choosing subsets of A and C to maximize the density of the subgraph. Finally, we build the dense tripartite graph G_3 out of the G_2 by adding all intermediate nodes $b \in B$ that are connected to at least one node of G_2.

In an ontology (see right inset of Figure 1), nodes from PO and GO are hierarchically arranged to reflect their relationships (e.g., is-a or part-of). The PAnG tool allows users to include restrictions on the ontology terms in the DSG. The simplest restriction is a *distance restriction* that specifies the maximal path length between pairs of nodes in set A (C). To this end, PAnG employs a distance metric d_A (d_C) and computes the densest subgraph G_3 that ensures that all node pairs of A (C) are within a given distance τ_A (τ_C). Furthermore, the user can filter the ontology by the relationship type, i.e., only node pairs that are in a specific relationship are considered for distance computation. The current version of PAnG [1] uses the simple *shortest path length* between a pair of terms as the distance metric. In this paper, we evaluate more sophisticated distance metrics in Section 4.

2.2   Graph Summarization

PAnG generates graph summaries for representing patterns. A summary of a tripartite annotation graph is also a graph. While there are many methods to summarize graphs, we focus on the graph summarization (GS) approach of [15]. Their graph summary is an aggregate graph comprised of a signature and corrections. It is the first application of minimum description length (MDL) principles to graph summarization and has the added benefit of providing intuitive coarse-level summaries that are well suited for visualization and link prediction.

A graph summary (GS) of a graph $G = ((A, B, C), (X, Y))$ consists of a graph **signature** $\Sigma(G)$ and a set of **corrections** $\Delta(G)$. The graph signature is defined as follows: $\Sigma(G) = ((S_{AC}, S_B), S_{XY})$. The sets S_{AC} and S_B are a disjoint partitioning of $A \cup C$ and B, respectively, that cover all elements of these sets. Each element of S_{AC} or S_B is a **supernode** and consists of one or more nodes of the original graph. An element of S_{XY} is a **superedge** and it represents edges between supernodes, i.e., $S_{XY} \subseteq S_{AC} \times S_B$. The **corrections** are the sets of edge additions and deletions $\Delta(G) = (S_{add}, S_{del})$. All edge additions are edges of the original graph G, i.e., $S_{add} \subseteq X \cup Y$. Deletions are edges between nodes of G that do not have an edge in the original graph, i.e., $S_{Del} \subseteq ((A \cup C) \times B) - (X \cup Y)$.

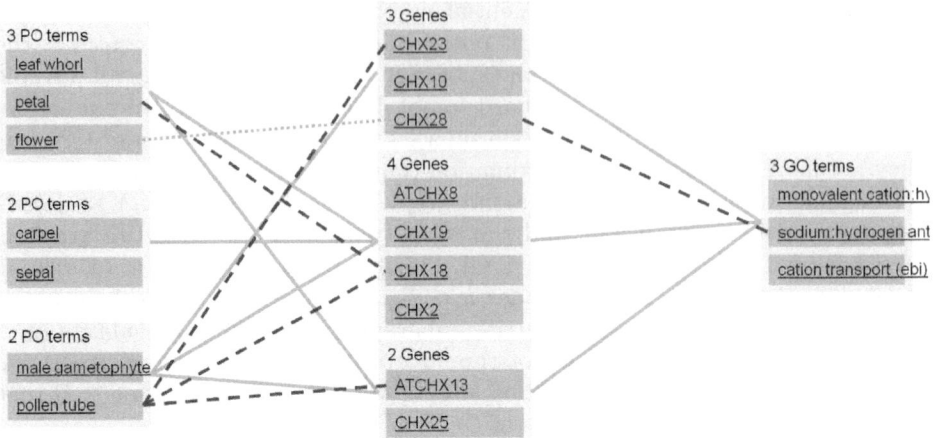

Fig. 3. Screenshot of a graph summary as generated by PAnG. Superedges are represented by green solid lines. Corrections include red dashed (deletion) and blue dotted (addition) lines, respectively.

Graph summarization is based on a two-part minimum description length encoding. The complexity of the original GS problem is currently unknown. However, if nodes are allowed to belong to more than one super node (i.e., overlapping supernodes), the problem reduces to finding the maximum clique in a graph, which is NP-hard. We use a greedy agglomerative clustering heuristic. The possible summaries of a graph will depend on the cost model used for an MDL encoding. In general, the cost model assigns weights to the number of superedges, deletions, and additions, respectively. Graph summarization looks for a graph summary with a minimal cost. Currently PAnG employs a simple cost model that gives equal weight to supernodes, superedges, and corrections.

2.3   Example Graph Summary

The Sze Laboratory at the University of Maryland is studying 20 CHX (Cation/H+ Exchanger) genes within the CPA2 family [23]; Figure 3 shows a dense subgraph (DSG) of the following 9 genes: CHX2, CHX10, CHX18, CHX19, CHX23, CHX25, CHX28, ATCH8, and ATCH13. A *supernode* (shaded rectangle) groups together either genes or terms from GO or PO.[1] For example, there are 3 gene supernodes; the top supernode includes 3 genes, CHX23, CHX10, and CHX28 while the middle supernode has 4 genes, ATCHX8, CHX19, CHX18, and CHX2. Figure 3 also includes 3 PO supernodes with 3, 3, and 2 terms, respectively; this summary also contains 1 GO supernode with 3 terms.

A *superedge* is a thick edge in the figure and occurs between 2 supernodes; it represents that all nodes in both the supernodes are connected to each other. For example, the superedge between the middle PO supernode with 2 PO terms

---

[1] Supernodes can also group dissimilar terms if desired, e.g., GO and PO terms.

carpel and sepal, and the middle gene supernode with 4 genes indicates that all 4 genes are each annotated with both PO terms.

We use the supernodes and superedges to explain the patterns. The top gene supernode (with 3 genes) has 1 superedge to the bottom PO supernode (with 2 terms). In contrast, the middle gene supernode (4 genes) has 3 superedges to each of the 3 PO supernodes. The bottom gene supernode (2 genes) has 2 superedges. Thus, the pattern distinguishes the 4 genes in the middle gene supernode, each annotated with 7 PO terms, from the 3 genes in the top gene supernode with the least number of PO annotations. CHX18 in the middle gene supernode is an outlier as will be discussed. 8 genes (except CHX28) are also annotated with all 3 GO terms in the GO supernode; thus, the gene function behavior of these 8 genes is identical with respect to these 3 GO terms. Sze confirmed the consistency of these patterns with results reported in [24].

Finally, the summary in Figure 3 illustrates *deletion* edges; these are broken edges in the figure and represent a deviation of behavior. A *deletion* reflects that a gene *does not have* a particular GO or PO annotation that is shared by the other genes (within the supernode). For example, CHX18 (middle supernode) is *not annotated* with PO terms petal or pollen tube; this is consistent with tissue localization results in [24]. CHX28 (top supernode) is not annotated with GO term sodium:hydrogen.... While this gene has not been studied, the patterns appears consistent with function based on phylogenetic tree analysis [23].

3 Preliminary Cross-Genome Validation

3.1 Data Collection Protocol and Statistics

Ideally, PAnG would use tools such as Inparanoid [10], Plaza [18], and Homologene [9] to find all known homologs of a given gene, in some alternate organism. For our proof-of-concept prototype, we apply a simpler protocol to identify genes with shared annotations. Collaborators Sze and Haag identified families of Arabidopsis or *C elegans* genes, respectively, as genes of interest. We then used GO terms describing their function to retrieve corresponding genes in a sister organism.

- **At_8 and Ce_9**
  At_8: Eight Arabidopsis genes in families labeled NHX or SOS; responsible for ion transport; seven are members of a sodium proton exchanger family.
  Ce_9: Nine *C. elegans* genes in families labeled nhx or pbo; all are members of a sodium proton exchanger family.
- **At_37 and Ce_53**
  At_37: 37 Arabidopsis genes. We started with a collection of 19 genes, identified by Sze, as occurring in family number(s) 212, 277, and 469; all are putative heavy ion transporting P2A-type ATPase genes [4,25]. We expanded this set to include all genes in families labeled ACE, ECA, HMA, RAN, and PAA.
  Ce_53: 53 *C. elegans* genes that are annotated with terms ion transport and/or divalent cations.

Table 1. Statistics for datasets At_8 and Ce_9. The 4 overlapping biological processes are cation transport, regulation of pH, sodium ion transport, and transmembrane transport. The 2 overlapping molecular functions are sodium:hydrogen antiporter activity and solute:hydrogen antiporter activity.

Dataset	Genes	Annotation	unique GO Terms	Biol. Proc.	Cell. Comp.	Molec. Funct.
At_8	8	117	28	17	6	5
Ce_9	9	91	25	20	3	2
Overlap			6	4	0	2

Table 2. Statistics for datasets At_37 and Ce_53. The 8 overlapping biological processes are ATP catabolic process, ATP biosynthetic process, transport, cation transport, metabolic process, response to metal ion, response to manganese ion, and manganese ion homeostasis. The 6 overlapping cellular components are intracellular, nucleus, cytoplasm, vacuolar membrane, membrane, and integral to membrane. The 4 overlapping molecular functions are zinc ion binding, coupled to transmembrane movement of ions, phosphorylative mechanism, and ATPase activity.

Dataset	Genes	Annotation	unique GO Terms	Biol. Proc.	Cell. Comp.	Molec. Funct.
At_37	455	37	106	55	29	22
Ce_53	53	685	48	21	9	18
Overlap			17	8	6	4

We next report on the number of (distinct) GO terms associated with each dataset, as well as the overlap, along the three GO branches, biological process (GO-BP), molecular function (GO-MF) and cellular component (GO-CC) in Table 1 and Table 2, respectively. We also report on the total number of annotations since multiple genes in the dataset could be annotated with the same GO term [2].

3.2   Cross-Genome Validation Using GS and DSG+GS Summaries

Figure 4 shows a graph summary (GS) for 8 genes in **At_8**. GO-BP terms are on the right and GO-MF and GO-CC on the left. Two gene supernodes include (NHX2, NHX6), and (NHX3, NHX4, NHX5), respectively. All the genes are annotated with the 3 GO-BP terms in the top GO-BP supernode. They do not appear to share many other GO-BP terms. Similarly, the 8 genes do not appear to share many GO-MF or GO-CC terms. We note that there is a deletion edge indicating that while NHX2 is associated with both sodium hydrogen antiporter and sodium ion transmembrane transporter function, NHX6 which shares many functions with NHX2 and is in the same gene supernode, is not annotated with sodium ion transmembrane transporter function.

---

[2] We further note that there are cases where a single gene is annotated more than once with the same GO term; this occurs when there is dissimilar annotation evidence from multiple sources.

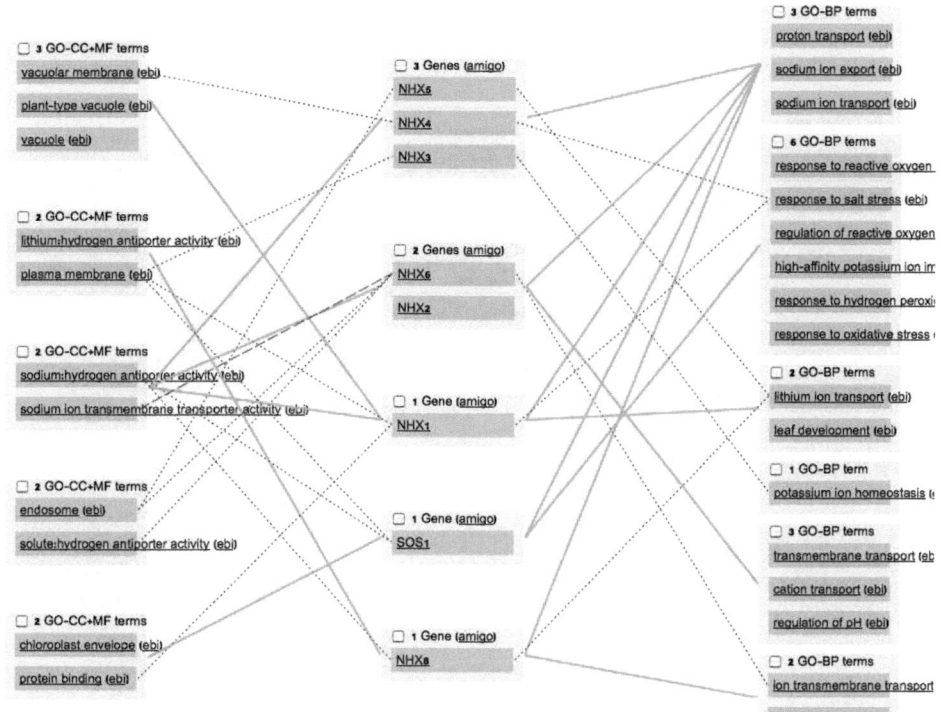

Fig. 4. Graph Summary (GS) of GO annotations for 8 genes in **At_8**; GO-BP on the right and GO-MF and GO-CC on the left

Figure 5 shows the graph summary (GS) for 9 genes in **Ce_9**. GO-BP terms are on the right and GO-MF and GO-CC on the left. All 9 genes share the three GO-MF terms in the supernode on the upper left. They also share 4 GO-BP terms grouped into a supernode on the right in the middle of the figure. 8 of the genes are gouped into a single gene supernode, with **nhx-2** being the outsider. This is because **nhx-2** appears to be much more richly annotated with an additional 11 GO-BP terms. Only **nhx-1** shares a single GO-BP term in that group, **positive regulation of growth rate**. However, these 8 genes in the supernode do not appear to share many additional GO-BP annotations. For example, only **nhx-1**, **nhx-4** and **nhx-6** share the GO-BP annotation **embryo development ending in birth**.

Finally, we combine the 8 Arabidopsis and the 9 C. elegans genes. We then identify a dense subgraph (DSG) with no distance restrictions. Figure 6 illustrates the benefit of creating a DSG prior to applying the graph summary (GS); DSG+GS identifies a *single* gene supernode that includes the 9 C. elegans genes and 2 Arabidopsis genes, **NHX2** and **NHX6**. These 2 Arabidopsis genes are included since both are annotated with the 4 GO-BP terms that annotated all 9 C. elegans genes as well as one (two) GO-MF terms. We note that the same 4

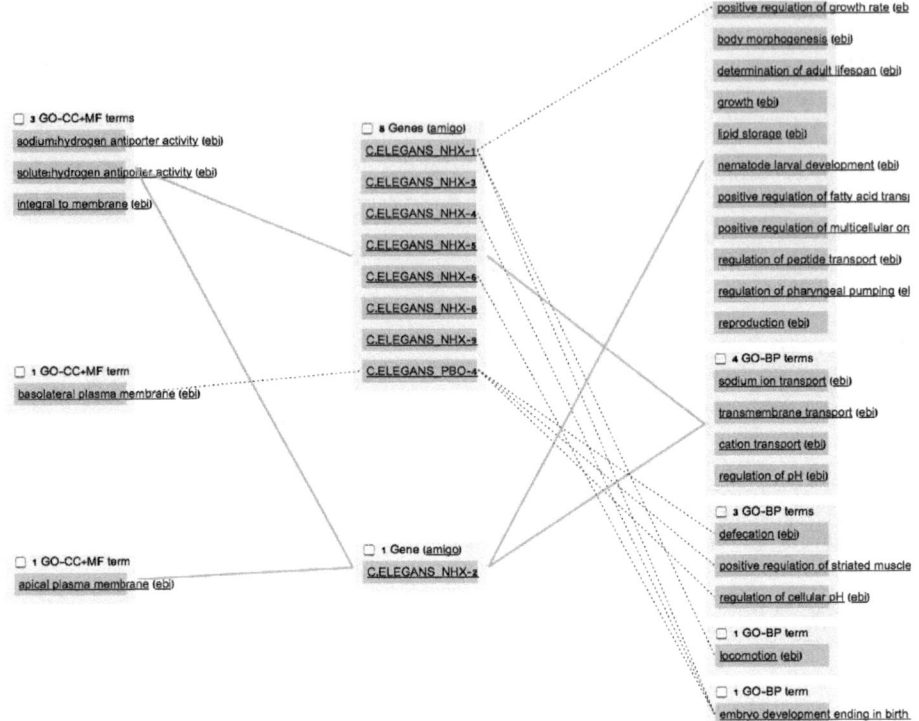

Fig. 5. Graph Summary (GS) of GO annotations for 9 genes in **Ce_9**; GO-BP on the right and GO-MF and GO-CC on the left

GO-BP terms and 2 of the 3 GO-MF of Figure 6 were identified in the overlap of Table 1. Thus, the DSG+GS annotation pattern is consistent with the shared annotations and overlap. Further, the DSG+GS annotation pattern provides a more detailed and nuanced understanding compared to the simple data of the overlap.

Based on phylogeny studies, NHX5 and/or NHX6 (intramembrane/Golgi phenotype) are more likely to be close homologs to *C. elegans* genes. NHX2 is part of the [NHX1-NHX4] group with a vacuolar-localized phenotype; phylogenetic studies show they are typically *plant-specific* genes. Thus, the supernode grouping of NHX2 and NHX6 with the 9 *C. elegans* genes appears to be partially validated using biological knowledge but requires further study to determine if this grouping may also have resulted from an incomplete annotations of these genes.

4   Using Distance Metrics for Validation

In Section 3, the dense subgraph (DSG) did not consider any distance restriction between the pairs of GO terms in the subgraph. Similar the graph summarization

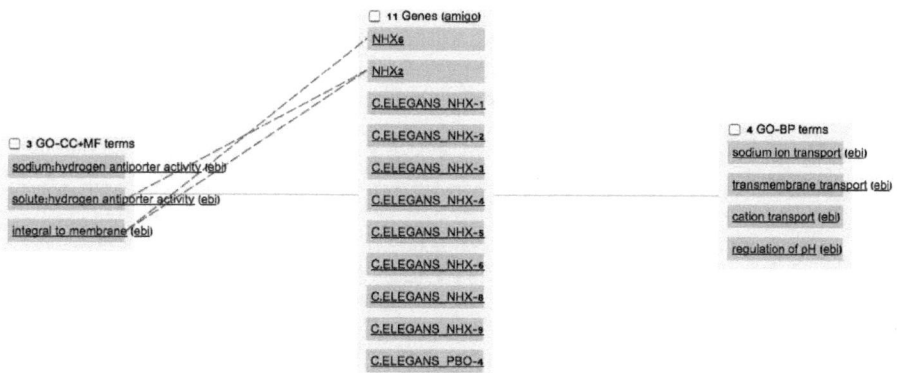

Fig. 6. Graph Summary of a Dense Subgraph (DSG+GS) of GO annotations for 8 genes in **At_8** and 9 genes in **Ce_9**; GO-BP on the right and GO-MF and GO-CC on the left

(GS) did not consider any distance restriction when constructing GO supernodes. This has significant limitations since terms in GO reflect concepts, and proximity (parents, siblings and neighbors) reflect relatedness of these concepts.

In this section, we consider two distance metrics that can be applied to taxonomies to measure the relatedness or similarity of concepts. We recognize that relatedness and similarity are not always synonymous. Our proposed metric is labeled d_{tax} and we compare it to d_{ps} [16], a state-of-the-art metric from the literature. We report on experiments performed on several datasets. In the range $[0.0 \ldots 1.0]$, where 1.0 represents no similarity, d_{tax} provides a wider dispersion of values, compared to d_{ps}. This wider dispersion provides d_{tax} with better discrimination of concepts that are not related, and hence more suited to the task of identifying related concepts.

4.1   Distance Metrics

The taxonomic organization of vertices in an ontology, as well as node properties such as descendants and ancestors, have been considered to develop state-of-the-art distance metrics that identify near neighbors, i.e., those that are proximal to each other in the taxonomy [11,14,17,20,28]. Consider the taxonomy of Figure 7(a). A *good* taxonomic distance metric should reflect that while the number of edges (say shortest path length) between a pair of nodes (1, 9) and (4, 17) may both be equal to 2, the taxonomic distances between these pairs should be different. The reasoning is that nodes that are deeper in the hierarchy and farther from the root are more specific.

Taxonomic distance metrics take values in the range $[0.0 \cdots 1.0]$, where 0.0 represents the greatest similarity. A desirable property is that two nodes that are (1) farther from the root and (2) closer to their lowest common ancestor, should be closer in distance. For example, in Figure 7(a), the pair of nodes (8,

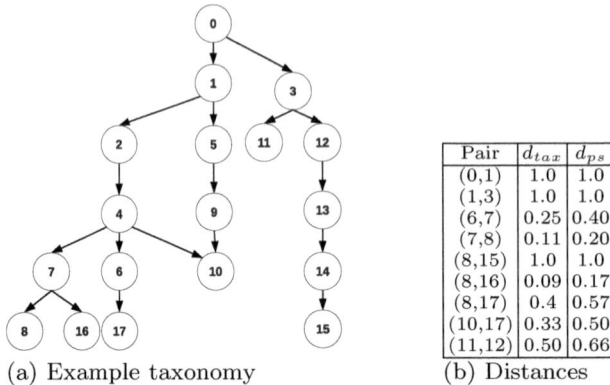

Pair	d_{tax}	d_{ps}
(0,1)	1.0	1.0
(1,3)	1.0	1.0
(6,7)	0.25	0.40
(7,8)	0.11	0.20
(8,15)	1.0	1.0
(8,16)	0.09	0.17
(8,17)	0.4	0.57
(10,17)	0.33	0.50
(11,12)	0.50	0.66

(a) Example taxonomy   (b) Distances

Fig. 7. Example of taxonomic distance(s) between pairs of vertices for an example taxonomy

16) should have a lower taxonomic distance to each other compared to the pair (11, 12), although the path length $= 2$ for both pairs. This is because the pair (8, 16) is farther from the root compared to (11, 12). The depth of a node from the root and the lowest common ancestor are defined as follows:

Definition 1 (Vertex Depth). *Given a directed graph G, the **depth** of a vertex x in G is the length of the longest path from the root of G to x.*

Definition 2 (Lowest Common Ancestor [3]). *Given a directed graph G, the **lowest common ancestor** of two vertices x and y, is the vertex of greatest depth in G that is an ancestor of both x and y.*

Let $d(x, y)$ be the number of edges on the longest path between vertices x and y in a given ontology. Also let $lca(x, y)$ be the lowest common ancestor of vertices x and y.

We propose a taxonomic distance (d_{tax}) which is defined as follows:

$$d_{tax}(x, y) = \frac{d(lca(x, y), x) + d(lca(x, y), y)}{d(root, x) + d(root, y)} \tag{1}$$

where *root* is the root node in the ontology.

We compare d_{tax} to a state-of-the-art distance metric d_{ps} [16] which is defined as follows:

$$d_{ps}(x, y) = 1 - \frac{d(root, lca(x, y))}{d(root, lca(x, y)) + d(lca(x, y), x) + d(lca(x, y), y)} \tag{2}$$

The intuition behind the d_{ps} metric proposed by Pekar and Staab [16] is that it captures the ability to represent the taxonomic distance between two vertices with respect to the depth of the common ancestor of these two vertices. Our proposed d_{tax} distance metric is to assign low(er) values of taxonomic distance to pairs of vertices that are (1) at greater depth in the taxonomy and (2) are

Table 3. Average and Variance of Taxonomic Distance(s) between GO-BP

Dataset	d_{tax}		d_{ps}	
	average	std_dev	average	std_dev
At_8 Arabidopis	0.936	0.207	0.964	0.133
Ce_9 C. elegans	0.868	0.275	0.925	0.182
At_8 ∩ Ce_9	0.872	0.270	0.925	0.185
At_8 ∪ Ce_9	0.962	0.159	0.977	0.106
At_37 Arabidopis	0.817	0.324	0.877	0.246
Ce_53 C. elegans	0.947	0.187	0.974	0.103
At_37 ∩ Ce_53	0.820	0.316	0.881	0.220
At_37 ∪ Ce_53	0.965	0.153	0.980	0.097

closer to their lowest common ancestor. Although d_{tax} distance metric satisfies theoretical distance properties, i.e., zero law, symmetry and triangle inequality, we do not focus on the formalization of these properties in this paper. In contrast, we show an empirical analysis of d_{tax} and how it compares to d_{ps} when both metrics are used to measure the relatedness or similarity of taxonomic concepts.

4.2 Properties of the Distance Metrics

Figure 7(b) illustrates the values assigned by both d_{tax} and d_{ps} to vertices in the taxonomy shown in Figure 7(a). In general, both metrics are able to assign values close to 0.0 to pairs of vertices separated by a small number of edges, and a value close to 1.0 to pairs of vertices separated by a large number of edges, e.g., (0, 1) and (8, 1). However, consider the pairs (10, 17) and (11, 12); d_{tax} is able to distinguish that both pairs have different taxonomic properties, i.e., the ratio $\frac{d_{tax}(10,17)}{d_{tax}(11,12)}$ is 0.6. Note that a value of 1.0 for this ratio implies that the taxonomic distances are judged to be similar. However, d_{ps} is not able to identify that these two pairs have different taxonomic properties. It assigns values such that the ratio $\frac{d_{ps}(10,17)}{d_{ps}(11,12)}$ is 0.75, i.e., closer to 1.0.

Next we report on the *distribution* of the pairwise distance d_{tax} and d_{ps} for several datasets (Table 3). We focus on the GO-BP terms and report on average and standard deviation. For **At_8** and **Ce_9**, the GO-BP terms in the intersection were more closely related compared to the individual datasets. The average distance for d_{tax} is also observed to be lower than the average for d_{ps}. Similarly, **Ar_37** had many pairs that were very close while there were also very distant pairs. We observe that the average is lower while there is also higher variance in the values.

To understand the discrimination capability of both metrics we "bucketize" pairs of GO-BP terms in U_1 and U_2 based on the length of the shortest path between them. Table 4 reports on the number of pairs, and the average and standard deviation for both metrics.

Table 4. Taxonomic Distance(s) between GO-BP for **At_8 ∪ Ce_9** and **At_37 ∪ Ce_53** (bucketized by path length)

	Path Length	#Pairs	d_{tax}		d_{ps}	
			average	std_dev	average	std_dev
At_8 ∪ Ce_9	1	51	0.15	0.09	0.31	0.12
	2	396	0.79	0.35	0.85	0.27
	3	2217	0.95	0.18	0.97	0.11
	4	4527	0.98	0.11	0.99	0.06
	5	1850	0.99	0.07	1.00	0.03
	6	254	1.0	0.0	1.0	0.0
	7	21	1.0	0.0	1.0	0.0
At_37 ∪ Ce_53	1	7	0.13	0.09	0.26	0.17
	2	63	0.74	0.35	0.84	0.23
	3	210	0.97	0.14	0.99	0.07
	4	372	0.98	0.10	0.99	0.04
	5	304	1.0	0.04	1.0	0.02
	6	103	1.0	0.03	1.0	0.02
	7	21	1.0	0.0	1.0	0.0

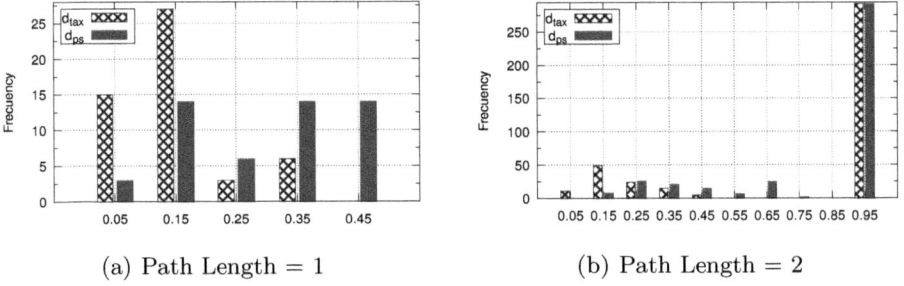

(a) Path Length = 1   (b) Path Length = 2

Fig. 8. Frequency Distributions of distance t_{tax} and d_{ps} for pairs of GO-BP terms in dataset **At_8 ∪ Ce_9** with Path Length = 1 and 2

We can observe that the d_{tax} average for path length = 1 and 2 are 0.15 and 0.79, whereas the values for d_{ps} are 0.31 and 0.85, respectively. This reflects that d_{tax} is more sensitive to path length, and is able to discriminate better than d_{ps}, when vertices are connected by a small number of edges. However, for distant vertices both metrics exhibit similar behavior.

Finally, we report on the *distribution* of values for d_{tax} and d_{ps}, for GO-BP terms in U_1, for path length = 1 and 2, in Figure 8. For path length = 1, 15 pairs have a value of 0.05 and 30 pairs have a value of 0.15, for d_{tax}. In contrast, 3 pairs have a value of 0.05 and 14 pairs have a value of 0.15, for d_{ps}.

To summarize, d_{tax} appears to be more sensitive in capturing the range of (dis)similarity or distance between pairs of terms. In contrast, d_{ps} appears to compress the distance distribution. Thus, d_{tax} appears to be more useful in differentiating closer pairs from more distant pairs.

5   Summary and Conclusions

We present a tool, PAnG (Patterns in Annotation Graphs), that is based on a complementary methodology of graph summarization and dense subgraphs. Our collaborators (Heven Sze an Arabidopsis specialist and Eric Haag who studies *C. elegans*) helped us to validate potential cross-genome annotation patterns using gene families with shared function in the two organisms. We demonstrate that a proposed metric for taxonomic distance d_{tax} is better able to discriminate among pairs of GO terms. In future work, we plan large scale cross-genome experiments and human validation of both the annotation patterns and the relatedness of pairs of GO terms. Further, we plan to study the properties of the proposed distance metric in ontologies such as MeSH [3] and Plant Ontology [4].

References

1. Anderson, P., Thor, A., Benik, J., Raschid, L., Vidal, M.E.: Pang - finding patterns in annotation graphs. In: Proceedings of the ACM Conference on the Management of Data (SIGMOD) (2012)
2. Ashburner, M., Ball, C.A., Blake, J.A., Botstein, D., Butler, H., Cherry, J.M., Davis, A.P., Dolinski, K., Dwight, S.S., Eppig, J.T., Harris, M.A., Hill, D.P., Issel-Tarver, L., Kasarskis, A., Lewis, S., Matese, J.C., Richardson, J.E., Ringwald, M., Rubin, G.M., Sherlock, G.: Gene Ontology: tool for the unification of biology. Natgenet 25(1), 25–29 (2000)
3. Bender, M.A., Farach-Colton, M., Pemmasani, G., Skiena, S., Sumazin, P.: Lowest common ancestors in trees and directed acyclic graphs. Journal of Algorithms 57(2), 75–94 (2005)
4. Bock, K., Honys, D., Ward, J., Padmanaban, S., Nawrocki, E., Hirschi, K., Twell, D., Sze, H.: Integrating membrane transport with male gametophyte development and function through transcriptomics. Plant Physiology 140(4), 1151–1168 (2006)
5. Charikar, M.: Greedy Approximation Algorithms for Finding Dense Components in a Graph. In: Jansen, K., Khuller, S. (eds.) APPROX 2000. LNCS, vol. 1913, pp. 84–95. Springer, Heidelberg (2000)
6. Garcia-Hernandez, M., Berardini, T.Z., Chen, G., Crist, D., Doyle, A., Huala, E., Knee, E., Lambrecht, M., Miller, N., Mueller, L.A., Mundodi, S., Rciser, L., Rhee, S.Y., Scholl, R., Tacklind, J., Weems, D.C., Wu, Y., Xu, I., Yoo, D., Yoon, J., Zhang, P.: TAIR: a resource for integrated Arabidopsis data. Functional and Integrative Genomics 2(6), 239 (2002)
7. Gene Ontology Consortium: The gene ontology project in 2008. Nucleic Acids Res. 36(Database Issue), D440–D444 (2008)
8. Goldberg, A.V.: Finding a maximum density subgraph. Tech. Rep. UCB/CSD-84-171, EECS Department, University of California, Berkeley (1984), http://www.eecs.berkeley.edu/Pubs/TechRpts/1984/5956.html

---

[3] http://www.nlm.nih.gov/mesh/
[4] http://www.plantontology.org/

9. Homologene, `http://www.ncbi.nlm.nih.gov/homologene`
10. Inparanoid, `http://inparanoid.sbc.su.se/cgi-bin/index.cgi`
11. Jiang, J., Conrath, D.: Semantic similarity based on corpus statistics and lexical taxonomy. CoRR cmp-lg/9709008 (1997)
12. Khuller, S., Saha, B.: On Finding Dense Subgraphs. In: Albers, S., Marchetti-Spaccamela, A., Matias, Y., Nikoletseas, S., Thomas, W. (eds.) ICALP 2009. LNCS, vol. 5555, pp. 597–608. Springer, Heidelberg (2009)
13. Lawler, E.: Combinatorial optimization - networks and matroids. Holt, Rinehart and Winston, New York (1976)
14. Lin, D.: An information-theoretic definition of similarity. In: ICML, pp. 296–304 (1998)
15. Navlakha, S., Rastogi, R., Shrivastava, N.: Graph summarization with bounded error. In: Proc. of Conference on Management of Data (SIGMOD) (2008)
16. Pekar, V., Staab, S.: Taxonomy learning - factoring the structure of a taxonomy into a semantic classification decision. In: COLING (2002)
17. Pesquita, C., Faria, D., Falcão, A., Lord, P., Couto, F.: Semantic similarity in biomedical ontologies. PLoS Computational Biology 5(7), e1000443 (2009)
18. Inparanoid, `http://bioinformatics.psb.ugent.be/plaza/`
19. Reiser, L., Rhee, S.Y.: Using The Arabidopsis Information Resource (TAIR) to Find Information About Arabidopsis Genes. Current Protocols in Bioinformatics, JWS (2005)
20. Resnik, P.: Using information content to evaluate semantic similarity in a taxonomy. In: IJCAI, pp. 448–453 (1995)
21. Rhee, S.Y., Beavis, W., Berardini, T.Z., Chen, G., Dixon, D., Doyle, A., Garcia-Hernandez, M., Huala, E., Lander, G., Montoya, M., Miller, N., Mueller, L.A., Mundodi, S., Reiser, L., Tacklind, J., Weems, D.C., Wu, Y., Xu, I., Yoo, D., Yoon, J., Zhang, P.: The Arabidopsis Information Resource (TAIR): a model organism database providing a centralized, curated gateway to arabidopsis biology, research materials and community. Nucleic Acids Res. 31(1), 224–228 (2003)
22. Saha, B., Hoch, A., Khuller, S., Raschid, L., Zhang, X.-N.: Dense Subgraphs with Restrictions and Applications to Gene Annotation Graphs. In: Berger, B. (ed.) RECOMB 2010. LNCS, vol. 6044, pp. 456–472. Springer, Heidelberg (2010)
23. Sze, H., Chang, C., Raschid, L.: Go and po annotations for cation/h+ exchangers. Personal Communication (2011)
24. Sze, H., Padmanaban, S., Cellier, F., Honys, D., Cheng, N., Bock, K., Conejero, G., Li, X., Twell, D., Ward, J., Hirschi, K.: Expression pattern of a novel gene family, atchx, highlights their potential roles in osmotic adjustment and k+ homeostasis in pollen biology. Plant Physiology 1(136), 2532–2547 (2004)
25. List of arabidopsis thaliana transporter genes on sze lab page, `http://www.clfs.umd.edu/CBMG/faculty/sze/lab/AtTransporters.html`
26. The Plant Ontology Consortium: The plant ontology consortium and plant ontologies. Comparative and Functional Genomics 3(2), 137–142 (2002), `http://dx.doi.org/10.1002/cfg.154`
27. Thor, A., Anderson, P., Raschid, L., Navlakha, S., Saha, B., Khuller, S., Zhang, X.-N.: Link Prediction for Annotation Graphs Using Graph Summarization. In: Aroyo, L., Welty, C., Alani, H., Taylor, J., Bernstein, A., Kagal, L., Noy, N., Blomqvist, E. (eds.) ISWC 2011, Part I. LNCS, vol. 7031, pp. 714–729. Springer, Heidelberg (2011)
28. Wang, J.Z., Du, Z., Payattakool, R., Yu, P.S., Chen, C.F.: A new method to measure the semantic similarity of go terms. Bioinformatics 23(10), 1274–1281 (2007)

Pay-as-You-Go Ranking of Schema Mappings Using Query Logs

Ruhaila Maskat*, Norman W. Paton, and Suzanne M. Embury

School of Computer Science, University of Manchester,
Manchester M13 9PL, United Kingdom
{maskatr,norm,embury}@cs.man.ac.uk

Abstract. Data integration systems typically make use of mappings to capture the relationships between the data resources to be integrated and the integrated representations presented to users. Manual development and maintenance of such mappings is time consuming and thus costly. Pay-as-you-go approaches to data integration support automatic construction of initial mappings, which are generally of rather poor quality, for refinement in the light of user feedback. However, automatic approaches that produce these mappings typically lead to the generation of multiple, overlapping candidate mappings. To present the most relevant set of results to user queries, the mappings have to be ranked. We proposed a ranking technique that uses information from query logs to discriminate among candidate mappings. The technique is evaluated in terms of how quickly stable rankings can be produced, and to investigate how the rankings track query patterns that are skewed towards specific sources.

Keywords: Schema Mapping, Ranking, Implicit Feedback, Dataspaces, Pay-as-you-go Data Integration.

1 Introduction

Data integration seeks to provide users with the illusion that there is a single data resource, containing well managed data, when in fact data is being obtained from several sources between which there may be heterogeneities of many types in the representations used. As a result, data integration is a tricky business, as the provision of an integrated representation involves the development of (potentially elaborate) mappings that seek to overcome the heterogeneities. The significant manual effort required to design and implement such mappings means that the construction of fully integrated resources is costly, giving rise to an interest in dataspaces, which support pay-as-you-go data integration [12].

In pay-as-you-go data integration, there are typically *bootstrapping* and *improvement* phases [14]. In the *bootstrapping* phase, an initial integration, typically of rather poor quality, is constructed automatically, building on schema matching [29], and generating mappings based on the matchings (e.g. [34]). In the

---

* Corresponding author.

O. Bodenreider and B. Rance (Eds.): DILS 2012, LNBI 7348, pp. 37–52, 2012.
© Springer-Verlag Berlin Heidelberg 2012

improvement phase, feedback of some form is obtained that enables mappings to be selected/discarded, ranked or refined. Typically the feedback is *explicit* – the user is asked for, or given the opportunity to provide, feedback (e.g., on the mappings directly [5] or on query results [3]). Explicit feedback provides evidence of a form that can be used directly to make well-informed decisions on the properties of different mappings, but its collection is intrusive in the sense that users have to devote time to its provision. As such, there is certainly value in considering situations in which *implicit* feedback can be used to improve an integration. Implicit feedback is collected as a by-product of the use of the integration system; in data integration, examples of implicit feedback include query logs and click-through patterns. There are a few examples of the use of implicit feedback to support data integration. For example, usage-based schema matching [10] identifies missing matches between schemas from join conditions in query logs.

In this paper, we propose a technique for ranking schema mappings on their relevance to user queries, making use of information from query logs. The work is motivated by the observation that: (i) automatic bootstrapping techniques typically generate multiple mappings as a result of false positive matches and the exploration of alternatives by mapping generation algorithms; and (ii) there may be many different online sources of related data, some of which are of greater relevance to users than others. For example, in the life sciences, functional data on an organism is often available in several different cross-species repositories (e.g. for microarrays this would include ArrayExpress [28], GEO [8] and the Stanford Microarray Database [7]) as well as organism-specific resources (e.g. the *Saccharomyces* Genome Database [11]).

The contributions of the paper are as follows: (i) a proposal for a technique that uses query logs to rank mappings; and (ii) an empirical evaluation of the technique in a life science case study involving microarray experiments. The rankings produced could have different uses, for example to select a subset of mappings for evaluating a user request, or to select an order of evaluation of mappings where results are produced incrementally for user inspection.

The remainder of the paper is structured as follows. Section 2 describes the technical background and the approach adopted. Section 3 reports on the results of an evaluation over microarray data sources. Section 4 describes related work on implicit feedback and ranking, and Section 5 presents some conclusions.

2   Proposed Ranking Approach

The problem we are trying to solve is the ranking of mappings in ways that reflect user requirements. This section presents an approach to the ranking of mappings using information from query logs, on the basis that an understanding of a user's information needs can be obtained from their queries [25]. In using information from query logs for ranking, we are interested in ranking the sets of multiple candidate mappings generated for each table in the global schema. In ranking the mappings, we then assume that instance-level data from query logs can be used to infer the interests of a user in subsets of the extents of the tables in the global schema. Such instance level data is represented by literals in equality comparisons in query *where* clauses.

The rankings of mappings are obtained by sorting the mappings based on scores. To calculate the scores of mappings, we relate the tuples generated from the mappings to the literals in queries from the query logs. We are aware that relevance judgments can only be made by a user, and thus that any ranking can only estimate relevance. To calculate the ranking score, we apply two scoring schemes: the widely-used Term Frequency/Inverse Document Frequency (TF·IDF) scheme from the information retrieval community, and a variant that counteracts a property of TF·IDF that may be seen as problematic when it is applied to mappings.

2.1   Term Frequency · Inverse Document Frequency

In information retrieval, Term Frequency · Inverse Document Frequency (TF·IDF) [25] is used to discriminate between documents by estimating how relevant their contents are to a set of terms in a search string. It combines two different weighting schemes to determine relevance, namely Term Frequency (TF) and Inverse Document Frequency (IDF). In TF/IDF, terms are viewed as having different levels of importance; some terms are weighted more while others are less. Term frequency is denoted as $tf(t, d)$ and defined as the number of occurrences of term t in document d. The Inverse Document Frequency (IDF) of a term t is defined as,

$$idf(t, D) = log \frac{|D|}{df(t, D)}$$

where $|D|$ is the total number of documents in the corpus D, and $df(t, D)$ is the number of documents that contain t:

$$df(t, D) = |\{d \in D : t \in d\}|$$

The purpose of inverse document frequency is to give additional weight to terms that are found in few documents (and thus that can discriminate between documents), and to play down terms that appear in many documents. The TF·IDF score for a term t is the product of the term frequency and the inverse document frequency, so for a search string containing the set of terms q, the TF·IDF score of a document d in corpus D is:

$$TF \cdot IDF(q, d) = \sum_{t \in q} tf(t, d) \times idf(t, D)$$

2.2   Scoring Mappings

To apply TF·IDF in our setting, we consider each schema mapping to be an unstructured document and use the terms extracted from query logs in place of the terms from a search string. We ignore the structure of mapping results on the basis that the use of a term in a query provides evidence of a user's interest

in the term in a way that may well apply in other settings. Thus, for example,
if a user expresses a query over a *gene* table for *S. pombe* genes, this may well
indicate that the user would also be interested in *S. pombe* proteins in proteomics
experiments in a different query.

Occurrences of the terms extracted from query logs are searched for within the
extent of each schema mapping to produce the TF·IDF score. Thus the TF·IDF
score of a mapping m in the context of query log l is:

$$TF \cdot IDFScore(l, m) = \sum_{t \in l} TF \cdot IDF(t, extent(m))$$

where $extent(m)$ is a document containing the result of evaluating the mapping
m, and l contains the bag of the literals obtained from equality conditions in the
query log.

From the above, it can be seen that TF·IDF supports the ranking of mappings
based on the number of occurrences of relevant terms, while taking into account
the fact that some terms are more discriminating than others. The significant
impact of the term frequency means that, all other things being equal, TF·IDF
prefers mappings with large extents to those with smaller extents. This can,
however, provide counterintuitive rankings. For example, assume that $Source_1$
contains 1000 values of which 100 are relevant, and that $Source_2$ contains 200
values of which 50 are relevant. If we assume that all terms are equally discriminating, then TF·IDF will rank $Source_1$ higher than $Source_2$, even though only
10% of the values in $Source_1$ are relevant whereas 25% of the values in $Source_2$
are relevant.

To take account of the fraction of the extent of a mapping that is relevant,
we now propose a variation on TF·IDF, as follows. Let $size(m)$ be the number
of values in m (i.e. the number of non-null values in the attributes of the tuples
of m). We define the *Size Normalised* (SN) TF·IDF as:

$$TF \cdot IDFScoreSN(l, m) = \sum_{t \in l} (\frac{TF \cdot IDF(t, extent(m))}{size(m)} * log(size(m)))$$

where $extent(m)$ is a document containing the result of evaluating the mapping
m, and l contains the bag of the literals obtained from equality conditions in the
query log. The multiplication by $log(size(m))$ has the effect of preferring larger
mappings where the percentages of relevant tuples are similar.

In the experiments, we provide results using both scoring schemes, without
making a case that either is better than the other – they simply prioritise different
features of the resulting mappings.

3   Evaluation

This section presents an evaluation of the approach described in the previous
section. The evaluation investigates two questions: the size of query log that is
required to support stable rankings, and the impact of skew towards specific

resources on rankings. Thus we do not directly empirically evaluate the suitability of the rankings produced, as there is already a substantial literature on ranking schemes such as TF·IDF [13,23,32,30,31].

3.1 Experimental Setup

This investigation was carried out using two publicly accessible life science databases: the Stanford Microarray[1] and ArrayExpress[2] databases. Data from experiments conducted on *Saccharomyces Cerevisiae* (bakers yeast) was collected as our sample. We used a Postgres DBMS[3] to create separate relations to represent the source relations of each data source, where two of the relations represent the *Experiment* and *Result* tables of Stanford Microarray database, and one relation represents the experiments in ArrayExpress.

An extra relation has been created to act as a global schema, for which ten schema mappings have been manually created, which are described in Table 1. Three of the mappings are *basic mappings*, which simply map the source relations to the global schema. The rest of the mappings are *refined mappings*, which are produced by combining the three *basic mappings* using the *union, join, difference, intersection* and *selection* operators [3]. This reflects the fact that tools that generate schema mappings, such as Clio [15], systematically explore alternative derivations. Based on one of the ways to refine mappings using the *difference* operator suggested in [3], we have created an auxiliary relation containing nameless genes. These genes, to which biologists have not given any names, go by their Open Reading Frame (ORF) identifier, e.g., *YHR089C*. This can be found in Mappings 6 and 7. The sizes of these mappings are presented in Table 2.

Because the relevance of a schema mapping is specific to a user's information need, the ground truth can be difficult to determine and cannot be generalised to every user. Hence, instead of evaluating our ranking approach based on the *correctness* of the ranking (with respect to which mapping should be placed higher in rank than another and the reason for this choice), evaluations explore the stability of the rankings as more feedback is obtained and the extent to which rankings track changes in the focus of queries towards individual sources.

3.2 Evaluation: Effects of Varying Query Log Size

We first set out to evaluate how different sizes of query log affect the ranking score to identify how much implicit feedback is needed to produce a stable ranking (i.e. how much feedback is required to produce the same ranking as would be produced with much more feedback).

We investigated this by generating 10 query log files. To simulate user queries that form the query logs, we have developed a log file generator, which extracts

---

[1] Stanford Microarray Database - http://smd.stanford.edu/

[2] ArrayExpress Database - http://www.ebi.ac.uk/arrayexpress/

[3] Postgress DBMS – http://www.postgresql.org/

Table 1. Schema mapping descriptions

Type	Mapping	Description
Basic	1	SELECT all experiments from Stanford Experiment table.
Basic	2	SELECT all experiments from Stanford Result table.
Basic	3	SELECT all experiments from ArrayExpress table.
Refined	4	JOIN tables Experiment and Result in Stanford and SELECT experiments of type 'limit'.
Refined	5	SELECT experiments of types 'limit', 'growth', 'genotype' and 'time' in ArrayExpress.
Refined	6	UNION result sets from Stanford (Mapping 4) and ArrayExpress (Mapping 5).
Refined	7	SELECT experiments in Stanford (Mapping 4) and perform a DIFFERENCE with a relation containing nameless genes.
Refined	8	SELECT experiments in ArrayExpress (Mapping 5) and DIFFERENCE with a relation of nameless genes.
Refined	9	UNION Mapping 7 (Stanford) and Mapping 8 (ArrayExpress).
Refined	10	SELECT experiments conducted on yeast gene located on the *Watson* strand from Mapping 9.

random values from the extent of the global schema, as described in Algorithm 1. This log file simulates literal values from equality conditions for queries over the global schema. The size of each log file is given by the number of values it contains. The smallest file contains 100 values, and for each consecutive file an increment of 100 was applied, until the largest file of 1000 values was produced. This approach can also be viewed as one in which each log file is a snapshot from a single growing log file at several points in time.

Results and Discussion. The results are presented in Fig. 1 for TF·IDF scoring. The following can be observed.

Table 2. Mapping sizes based on total number of terms

Mapping	Size
1	18,047
2	36,561
3	106,876
4	199,737
5	75,379
6	255,940
7	182,213
8	73,031
9	255,261
10	126,052

Algorithm 1. Query log file generator for Experiment 1

1: Create log file *log* if it does not exist
2: **for** $i = 1 \rightarrow 100$ **do**
3:    *columnname* = select random column from global schema
4:    *tuplelist* = get all tuples from global schema where column name equals to
      columnname
5:    *value* = select random tuple from *tuplelist*
6:    append *value* into *log*
7: **end for**

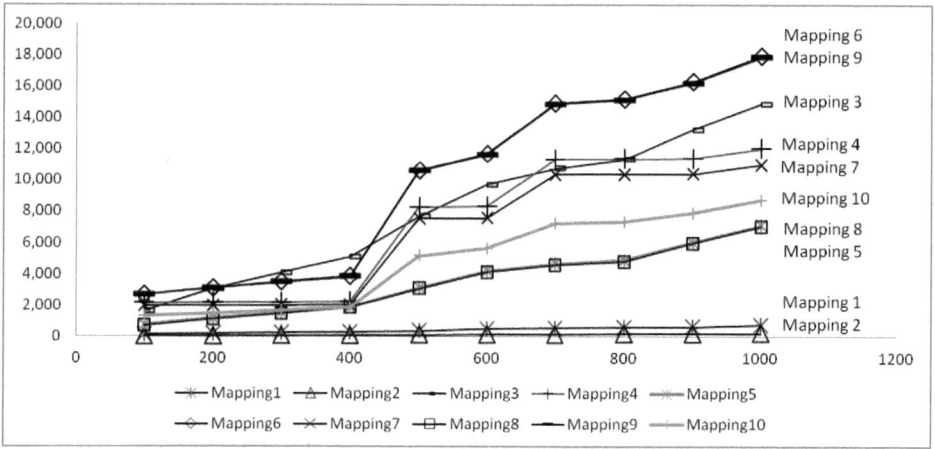

Fig. 1. TF·IDF ranking score for all mappings

- The TF·IDF scores for all mappings increase broadly linearly with the increase in the size of the query logs, reflecting the fact that the total score
  for each mapping is a sum of the scores for each term in the log.
- The ranked order of the mappings is broadly preserved across all the different
  log sizes, and from query log sizes of 500 and upwards is stable other than
  within groups of rankings that have similar overall scores. This shows that
  the rank order has been stable from quite small query logs consisting of
  randomly chosen terms.
- The sharp increase in the TF·IDF score for Mappings 6 and 9 from query
  log size 400 to 500 is due to the random inclusion of a term in the query
  log that happens to have a large number of occurrences in the extent of
  these mappings; similar features explain other sharp increases. It is a feature
  of most of the mappings that they contain many terms that have a low
  frequency and a few terms that have a high frequency, and thus that give
  rise to high term frequency scores.
- The TF·IDF scores of some mappings are much higher than others. This is
  explained principally by term frequency, which is correlated with mapping
  size.

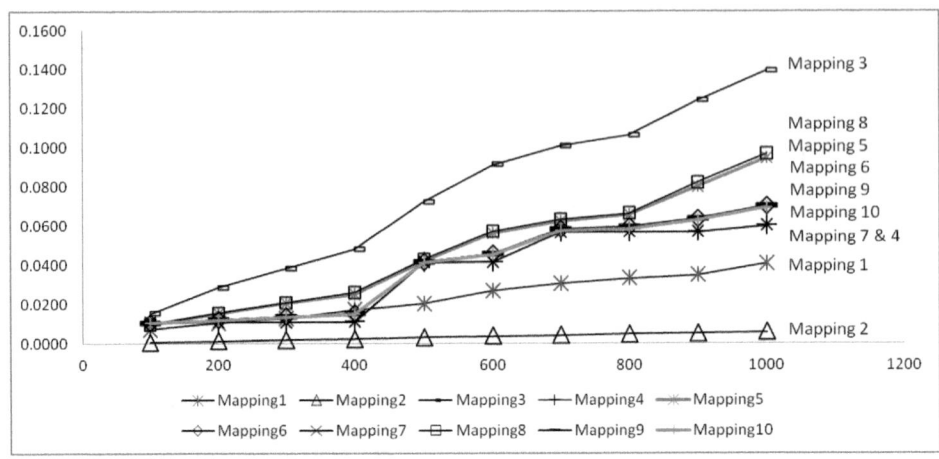

Fig. 2. Size normalised TF·IDF ranking scores for different log size

The results are presented in Fig. 2 for size normalized TF·IDF scoring. The following can be observed.

- Scores for size normalized TF·IDF of all mappings increase broadly linearly with the increase in the size of the query logs. The change in rankings of mappings indicates, for example, that although Mapping 3 has fewer relevant terms, the fraction of these terms in relation to Mapping 3's size of 106,876 is higher than for Mapping 6, which has more relevant terms and a size 255,940.
- Similar to the un-normalized experiment, the ranked order of the mappings is broadly preserved across all the different log sizes, although ranking stability is reached slightly earlier at log sizes of 400. This too shows that the rank order has been stable from quite small query logs consisting of randomly chosen terms.
- Sharp increases in the size normalized TF·IDF score are also present and are due to the same reasons observed in non-normalized TF·IDF.

3.3   Evaluation: Effects of Varying Query Log Skew

In ranking mappings according to their relevance to a user, the objective is to enable an information integration system to make informed decisions as to which mappings best meet user needs. Given that user queries may contain values that occur more frequently in some sources than others, this experiment investigates the extent to which such a skewed distribution, represented in query logs, is reflected in the rankings produced for mappings.

We define the skew of a query log l in relation to a source s to be the fraction of the queries in l that feature terms that are exclusive to s:

$$Skew(l, s) = \frac{N_{l_s}}{N_l}$$

Algorithm 2. Log File Generator for Skewed Logs

1: **Input:** *skew1* integer
2: **Input:** *skew2* integer
3: Create log file *log*
4: **for** $i = 1 \rightarrow skew1$ **do**
5:     *valuelist* = get all values from the global schema produced by (Mapping 1 or Mapping 2) but not Mapping 3
6:     *value* = choose a value at random from *valuelist*
7:     append *value* into *log*
8: **end for**
9: **for** $i = 1 \rightarrow skew2$ **do**
10:     *valuelist* = get all values from the global schema produced by Mapping 3 but not (Mapping 1 or Mapping 2)
11:     *value* = choose a value at random from *valuelist*
12:     append *value* into *log*
13: **end for**

---

where N_{l_s} is the number of queries in the query log that reference only terms in s and N_l is the total number of queries in the log.

Using the log file generator in Algorithm 2, 21 query logs were produced. Each log file contains 500 queries containing a controlled number of terms that are exclusive to each data source. For example, a query log may contain 50 ArrayExpress-exclusive terms (skew of 10%) and 450 Stanford-exclusive terms (skew of 90%). The level of skew is varied across the log files at intervals of 5%. As such, we started with a log file having 100% Stanford-exclusive queries, continued with a log file of 5% ArrayExpress-exclusive queries and 95% Stanford-exclusive queries, and ended with a log file containing 100% ArrayExpress-exclusive queries.

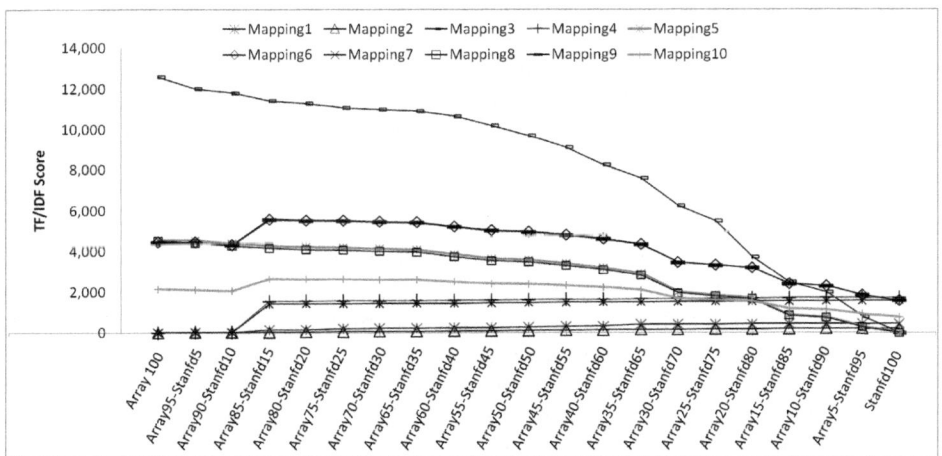

Fig. 3. TF·IDF scores for different levels of skew

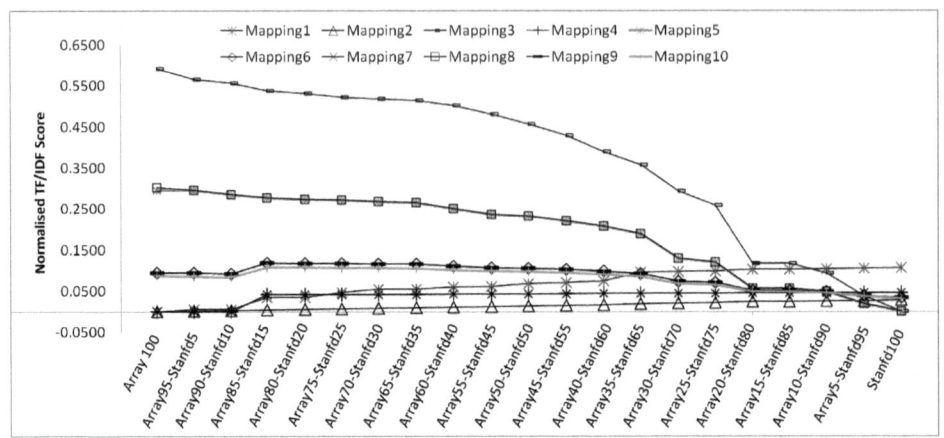

Fig. 4. Size normalised TF·IDF scores for different levels of skew

Results and Discussion. The results for TF·IDF are presented in Fig. 3 and Table 3, and for Size Normalised TF·IDF are presented in Fig. 4 and Table 4. The following can be observed:

– Scores for all three basic mappings (i.e., 1, 2 and 3) increase as the skew towards their specific data source rises, and this trend is followed by the source-specific refined mappings.
– The ranks for the neutral mappings tend to be fairly stable while the skew varies, as indicated in Tables 3 and 4.
– The rankings in the 50%-50% skew case are not the same as in Experiment 1 because the query logs now satisfy the property that they contain only terms that are specific to one source or another (i.e., no terms are found in both sources), and thus the values they contain are different from those used in Experiment 1 (in essence, terms that are common to different sources have been excluded). Note that the absence of terms that belong to both sources in this experiments has resulted in lower term frequencies for the neutral mappings in this experiment than in Experiment 1.
– Mapping 3 (ArrayExpress-exclusive) is ranked first even when the query logs are notably skewed toward Stanford (e.g., Stanford 80%). This reflects the fact that the scores for the mappings in the experiment cover a wide range of values, and thus that significant skew may be required to counteract other factors that contribute to a score (such as mapping size or term frequency distributions). This is not a problem; it simply reflects the fact that the mappings have significantly different properties. Mapping 3 is ranked first even when there is significant skew towards Stanford because: (i) in the case of TF·IDF it has the greatest quantity of relevant data, taking into account that some terms are more discriminating than others; and (ii) in the case

Table 3. Ranks obtained using TF·IDF scores

Mapping	Exclusivity	100% Array Express	50% Stanford 50% ArrayExpress	100% Stanford
1	Stanford	7	9	6
2	Stanford	7	10	7
3	ArrayExpress	1	1	8
4	Stanford	7	7	1
5	ArrayExpress	2	4	8
6	Neutral	4	2	2
7	Stanford	7	8	3
8	ArrayExpress	3	5	8
9	Neutral	5	3	3
10	Neutral	6	6	5

Table 4. Ranks obtained using normalised TF·IDF scores all mappings

Mapping	Exclusivity	100% Array Express	50% Stanford 50% ArrayExpress	100% Stanford
1	Stanford	7	7	1
2	Stanford	7	10	7
3	ArrayExpress	1	1	8
4	Stanford	7	8	2
5	ArrayExpress	3	2	8
6	Neutral	4	4	5
7	Stanford	7	9	3
8	ArrayExpress	2	3	8
9	Neutral	5	5	5
10	Neutral	6	6	6

of Size Normalised TF·IDF, Mapping 3 has the highest fraction of relevant data, taking into account that some terms are more discriminating than others.

– The size normalization has not had a substantial effect on the rankings of mappings across sources in this experiment. For example, the four Stanford exclusive mappings, Mappings 1, 2, 4 and 7, are ranked as the bottom four for the 50%-50% skew levels using both scoring functions. This is a feature of the data sets used in this experiment, and should not be assumed to apply in other contexts.

4   Related Work

4.1   Implicit Feedback

Several forms of implicit feedback have been proposed. Oard and Kim [27] suggest a framework that categorises user's *observable behaviours* during their activities of searching and using information in terms of their potential as a source of implicit feedback. These behaviours include *viewing, listening, printing, copying-and-pasting, quoting* and *annotating*. This has been further extended by Kelly and Teevan [21] to include *scrolling, finding, querying, typing* and *editing*.

Implicit feedback has been used for different purposes. For example, Elmele-egy *et al.* [9] used query logs to match database schemas, Joachims *et al.* [19] employed navigation patterns to infer a user's judgment on the relevance of certain web documents, Navarro *et al.* [26] suggested public collections of user tags to improve the relevance scores of web documents, and Kelly and Belkin [20] use the length of time a web document is displayed in the browser as an indication of a user's preferences.

In principle, all of the above approaches could be applied in dataspaces. In the dataspace literature, several types of implicit feedback have been described, though not involving query logs. Roomba [18] uses the rate of user clicking to indicate a user's level of interest in an answer, and a series of consecutively posed queries may be taken as implying that the terms found in successive queries describe related information needs. In PayGo [24], implicit feedback in the form of user clicks on answers is used to inform the choice of data sources and to improve the ranking of returned answers.

4.2   Relevance Ranking

Many ranking strategies have been developed in the database and information retrieval communities, where they largely try to prioritise query answers based on their relevance to a user query. A significant difference between ranking strategies in the two areas is that in databases, structure can be perceived as helpful in ascertaining the relevance of term. For example, a term "python" found in a column named *animal* is not likely to be relevant to a user who is searching for books about the Python programming language. By contrast, in information retrieval, text documents are the primary focus, and therefore the location of a term is not so straightforward to exploit. In this context, an answer is relevant if a user views it as fulfilling that user's information need [25]. Ranking is commonly seen as a process whose goal is to discriminate between answers, and thus indirectly to reduce the number of answers presented. Since only a user can judge the correctness of an answer, ranking in many cases can only provide a suggestion of possible relevance.

Ranking efforts for structured data can be separated into two groups with regard to the type of query performed. One group utilises keyword queries while another manipulates structured queries. Proposals that support keyword queries over a structured database (e.g., [16,17,2,4]) generally represent the underlying structured database using some form of graph-based model, obtain their terms from the user's query, and search for tuples containing the terms by traversing the constructed graph. The path that joins these tuples from multiple tables is generally known as a *tuple tree* or *join tree*. To rank answers, the tuple trees must first be ranked, commonly, by calculating the number of joins that make the tree. Fewer joins suggest higher relevance on the premise that more joins may have added information that does not contribute to the relevance of an answer [17]. Although our work assumes the set of terms from query logs are used in a similar fashion to a free-form keyword search, we abstract over the underlying structured database.

Where structured queries are used, the goal is generally to automatically rank query result tuples in a way that reflects a user's information need. Lange *et al.* [22] rank query results using a function that maps tuples as vectors to a manually curated feature model. A tuple having most of the desirable features would be ranked higher. Since in the life sciences the interpretation of concepts may differ among biologists, we find the construction of a feature model to be out of sync with our pay-as-you-go setting. Chaudhuri *et al.* [6] use a technique from information retrieval to calculate, from a query workload and current data, their dependencies and correlations. This information is then used to rank result tuples according to the probability that they satisfy a given query condition. We differ in that we do not assume prior knowledge of data dependencies and correlations.

Like us, Agrawal *et al.* [1] use TF·IDF to rank queries, although their context and approach differ. Their objective is to identify which queries from logs may have given rise to security breaches, where it is known what data has been leaked. Their approach compares the results of queries in the logs with the data that have been leaked, using a distance function that adopts principles from TF·IDF. Thus Agrawal *et al.* compare the extents of tables, using available structural information, whereas we relate mapping extents as unstructured resources to query conditions rather than their results.

5   Conclusions

Pay-as-you-go data integration aspires to bring the benefits of data integration to settings where more labour-intensive manual integration is not a practical proposition. This tends to mean domains in which there are rapidly changing or diverse requirements, and where there are numerous data resources. The form of payment in pay-as-you-go systems is clearly important; explicit feedback, for example on query results, has been shown to support a range of different improvements to emerging integrations (e.g. [33,3]). However, the collection of explicit feedback is intrusive, and thus maximum benefit must be derived from implicit feedback.

This paper has investigated the use of one form of implicit feedback from query logs for ranking mappings. Ranking of mappings has several applications, such as reducing the cost of evaluating integration queries, and supporting the timely provision of results that are expected to meet user needs. Ranking can also be used in conjunction with manual integration to direct manual efforts towards mappings for which there is evidence of relevance to users. Evaluations suggest that: (i) stable rankings can be produced even using query logs containing a few hundred queries; and (ii) that where user queries reflect requirements that are skewed towards specific sources, this is reflected in mapping rankings produced using TF·IDF. The subtleties reflected in the resulting rankings mean that there may be value in future work exploring alternative scoring functions to underpin rankings for structured mappings.

References

1. Agrawal, R., Evfimievski, A., Kiernan, J., Velu, R.: Auditing disclosure by relevance ranking. In: Proceedings of the 2007 ACM SIGMOD International Conference on Management of Data, pp. 79–90. ACM (2007)
2. Agrawal, S., Chaudhuri, S.: DBXplorer: A system for keyword-based search over relational databases. In: Data Engineering, 2002, pp. 5–16 (2002)
3. Belhajjame, K., Paton, N.W., Embury, S.M., Fernandes, A.A.A., Hedeler, C.: Feedback-based annotation, selection and refinement of schema mappings for dataspaces. In: Proceedings of the 13th International Conference on Extending Database Technology, EDBT 2010, pp. 573–584. ACM, New York (2010)
4. Bhalotia, G., Hulgeri, A., Nakhe, C.: Keyword searching and browsing in databases using BANKS. In: Data Engineering (2002)
5. Cao, H., Qi, Y., Selçuk Candan, K., Sapino, M.L.: Feedback-driven result ranking and query refinement for exploring semi-structured data collections. In: Proceedings of the 13th International Conference on Extending Database Technology, EDBT 2010, pp. 3–14. ACM, New York (2010)
6. Chaudhuri, S., Das, G., Hristidis, V., Weikum, G.: Probabilistic information retrieval approach for ranking of database query results. ACM Trans. Database Syst. 31(3), 1134–1168 (2006)
7. Demeter, J., et al.: The stanford microarray database: implementation of new analysis tools and open source release of software. Nucleic Acids Research 35(Database-Issue), 766–770 (2007)
8. Edgar, R., Domrachev, M., Lash, A.E.: Gene expression omnibus: Ncbi gene expression and hybridization array data repository. Nucleic Acids Research 30(1), 207–210 (2002)
9. Elmeleegy, H., Elmagarmid, A., Lee, J.: Leveraging query logs for schema mapping generation in U-MAP. In: Proceedings of the 2011 International Conference on Management of Data, SIGMOD 2011, pp. 121–132. ACM, New York (2011)
10. Elmeleegy, H., Ouzzani, M., Elmagarmid, A.: Usage-Based Schema Matching. In: International Conference on Data Engineering, pp. 20–29 (2008)
11. Engel, S.R., et al.: *Saccharomyces* genome database provides mutant phenotype data. Nucleic Acids Research 38(Database-Issue), 433–436 (2010)
12. Franklin, M., Halevy, A., Maier, D.: From databases to dataspaces: a new abstraction for information management. SIGMOD Rec. 34(4), 27–33 (2005)
13. Gospodnetic, O., Hatcher, E.: Lucene in Action (In Action series). Manning Publications (December 2004)

14. Hedeler, C., Belhajjame, K., Fernandes, A.A.A., Embury, S.M., Paton, N.W.: Dimensions of Dataspaces. In: Sexton, A.P. (ed.) BNCOD 2009. LNCS, vol. 5588, pp. 55–66. Springer, Heidelberg (2009)
15. Hernández, M.A., Miller, R.J., Haas, L.M.: Clio: A semi-automatic tool for schema mapping. In: SIGMOD Conference, p. 607 (2001)
16. Hristidis, V., Gravano, L., Papakonstantinou, Y.: Efficient IR-style keyword search over relational databases. In: Proceedings of the 29th International Conference on Very Large Data Bases, VLDB 2003, vol. 29, pp. 850–861. VLDB Endowment (2003)
17. Hristidis, V., Papakonstantinou, Y.: Discover: keyword search in relational databases. In: Proceedings of the 28th International Conference on Very Large Data Bases, VLDB 2002, pp. 670–681. VLDB Endowment (2002)
18. Jeffery, S.R., Franklin, M.J., Halevy, A.Y.: Pay-as-you-go user feedback for dataspace systems. In: Proceedings of the 2008 ACM SIGMOD International Conference on Management of Data, SIGMOD 2008, pp. 847–860. ACM, New York (2008)
19. Joachims, T., Granka, L., Pan, B., Hembrooke, H., Gay, G.: Accurately interpreting clickthrough data as implicit feedback. In: Proceedings of the 28th Annual International ACM SIGIR Conference on Research and Development in Information Retrieval, SIGIR 2005, pp. 154–161. ACM, New York (2005)
20. Kelly, D., Belkin, N.J.: Display time as implicit feedback: understanding task effects. In: Proceedings of the 27th Annual International ACM SIGIR Conference on Research and Development in Information Retrieval, pp. 377–384. ACM (2004)
21. Kelly, D., Teevan, J.: Implicit feedback for inferring user preference: a bibliography. SIGIR Forum 37(2), 18–28 (2003)
22. Lange, M., Spies, K., Bargsten, J., Haberhauer, G., Klapperstück, M., Leps, M., Weinel, C., Wünschiers, R., Weissbach, M., Stein, J., Scholz, U.: The LAILAPS search engine: relevance ranking in life science databases. J. Integr. Bioinform. 7(2), 110 (2010)
23. Lu, Z., Kim, W., John Wilbur, W.: Evaluating relevance ranking strategies for MEDLINE retrieval. Journal of the American Medical Informatics Association: JAMIA 16(1), 32–36 (2009)
24. Madhavan, J., Jeffery, S.F., Cohen, S., Dong, X., Ko, D., Yu, C., Halevy, A.: Web-scale data integration: You can only afford to pay as you go. In: Proceedings of CIDR, pp. 342–350 (2007)
25. Manning, C.D., Raghavan, P., Schtze, H.: Introduction to Information Retrieval. Cambridge University Press, New York (2008)
26. Bullock, B.N., Jäschke, R., Hotho, A.: Tagging data as implicit feedback for learning-to-rank. In: Proceedings of the ACM WebSci 2011, Koblenz, Germany, June 14-17, pp. 1–4 (2011)
27. Oard, D.W., Kim, J.: Modeling information content using observable behavior. Science, 481–488 (2001)
28. Parkinson, H.E., et al.: Arrayexpress update - an archive of microarray and high-throughput sequencing-based functional genomics experiments. Nucleic Acids Research 39(Database-Issue), 1002–1004 (2011)
29. Rahm, E., Bernstein, P.A.: A survey of approaches to automatic schema matching. The VLDB Journal 10, 334–350 (2001)
30. Salton, G., Waldstein, R.K.: Term relevance weights in on-line information retrieval. Information Processing and Management 14(1), 29–35 (1978)
31. Schlieder, T., Meuss, H.: Querying and ranking xml documents. Journal of the American Society for Information Science and Technology 53(6), 489–503 (2002)

32. Sugiyama, K., Hatano, K., Yoshikawa, M., Uemura, S.: Refinement of tf-idf schemes for web pages using their hyperlinked neighboring pages. In: Proceedings of the Fourteenth ACM Conference on Hypertext and Hypermedia, HYPERTEXT 2003, pp. 198–207. ACM, New York (2003)
33. Talukdar, P.P., Jacob, M., Mehmood, M.S., Crammer, K., Ives, Z.G., Pereira, F., Guha, S.: Learning to create data-integrating queries. Proc. VLDB Endow. 1(1), 785–796 (2008)
34. Xu, L., Embley, D.W.: A composite approach to automating direct and indirect schema mappings. Inf. Syst. 31(8), 697–732 (2006)

Combining Structured and Unstructured Knowledge Sources for Question Answering in Watson

Ken Barker

IBM Research, Thomas J. Watson Research Center,
Yorktown Heights, NY 10598, USA
kjbarker@us.ibm.com

1 Open-Domain Question Answering

One of the classical challenges of Artificial Intelligence research has been to build automatic, open-domain question answering (QA) systems. The goal is not merely to retrieve documents containing answers to questions, or to query databases known to contain the answers. Rather, open-domain question answering systems must accept any question on any topic, find relevant information from possibly disparate sources, synthesize an answer, explain the evidence supporting the answer and provide an indication of the systems confidence that the answer is correct.

In accepting questions from any domain, a QA system cannot rely on knowledge of select, curated information sources. It cannot assume that answers will come from predictable sources or that sources can be joined, integrated or otherwise combined in predictable ways to synthesize answers.

The open-domain task precludes complete, prior manual alignment of information sources. Some preemptive integration of domain-independent sources may be useful, but the QA system must be able to integrate information "on-the-fly" to support answer generation and explanation.

2 Watson at Play

Building on significant advances in open-domain QA [1–3], the DeepQA team at IBM Research committed to test its QA technology head-to-head against human question answerers at the game of *Jeopardy!*, an American television game show. In *Jeopardy!*, three contestants compete at answering cleverly worded questions over a wide range of topics. A contestant must understand the question, determine if she knows the answer and press a button before the other two contestants to win the opportunity to answer the question. Each question has an associated monetary value that is added to the contestants earnings for correct answers or subtracted from the contestants earnings for incorrect answers. In 2011, IBMs *Watson* QA system competed on the television show and beat the two best *Jeopardy!* champions over a two-game match.

O. Bodenreider and B. Rance (Eds.): DILS 2012, LNBI 7348, pp. 53–55, 2012.

Watson's success was due in large part to an architecture [4] that allows it to query many disparate sources (both structured sources and unstructured text), generate candidate answer hypotheses, find supporting evidence for the hypotheses among different information sources, combine and weigh the evidence from different sources, and merge and rank evidence-supported answer candidates. Many different hypotheses are generated and evidence for them explored in parallel, with statistical machine learning providing models of how to combine the information and determine if candidate answer confidence is strong enough to answer the question [5, 6]. *Watson* does not explicitly integrate structured information sources prior to QA. Rather, facts are merged during QA if text supporting them and models that use them suggest that doing so will improve accuracy.

3   Watson Goes to School

Demonstrating human-level performance at the *Jeopardy!* task represents a milestone in open-domain QA, and components in *Watson* have exceeded state-of-the-art benchmarks in several traditional Natural Language Processing test sets (parsing, entity disambiguation, relation detection and textual entailment) [4]. Nevertheless, it is still just one very well-constrained task.

Moving forward, we are expanding the DeepQA scientific process and technology to other applications, to specific domains and to more interactive use cases. *Watson 2.0* will provide interactive decision support, combining search and curated formal knowledge sources in flexible ways, and providing not just answers to questions, but considering larger problem scenarios and producing hypothetical solutions with domain-sensible explanations.

The healthcare / medicine domain is particularly interesting for exploring deeper integration of unstructured text and formal reasoning because of the wealth of material in textbooks and journals as well as curated, structured sources. Yet none of these materials on their own contain all of the necessary information to solve typical problems in the domain. For example, it is unlikely that a given text or structured resource is sufficient for determining the *most likely* diagnoses given a specific patients symptoms, medical history and demographics. Generating hypotheses from sources that drive exploration of evidence in other sources, and learning how to combine hypotheses and evidence allows sources to "fill each others gaps" without prior, complete integration. *Watson*s approach of using diverse structured and unstructured information sources to hypothesize solutions, and combining these with different kinds of evidence according to learned models shows great promise for advancing knowledge-based tools in this rich domain.

References

1. Maybury, M.: New Directions in Question-Answering. AAAI Press, Menlo Park (2004)
2. Strzalkowski, T., Harabagiu, S.: Advances in Open-Domain Question-Answering. Springer, Berlin (2006)

3. Ferrucci, D., Nyberg, E., Allan, J., Barker, K., Brown, E., Chu-Carroll, J., Ciccolo, A., Duboue, P., Fan, J., Gondek, D., Hovy, E., Katz, B., Lally, A., McCord, M., Morarescu, P., Murdock, B., Porter, B., Prager, J., Strzalkowski, T., Welty, C., Zadrozny, W.: Towards the Open Advancement of Question Answering Systems. IBM Research Report RC24789. IBM, Armonk, New York (2009)
4. Ferrucci, D.: Introduction to This is Watson. In: Murdock, W.J. (ed.) IBM Journal of Research and Development, vol. 56(3/4), pp. 1:1–1:15. IBM, New York (2012)
5. Ferrucci, D., Brown, E., Chu-Carroll, J., Fan, J., Gondek, D., Kalyanpur, A., Lally, A., Murdock, J.W., Nyberg, E., Prager, J., Schlaefer, N., Welty, C.: Building Watson: An Overview of the DeepQA Project. AI Magazine 31(3), 59–79 (2010)
6. Murdock, W.J. (ed.): This is Watson. IBM Journal of Research and Development, vol. 56(3/4). IBM, New York (2012)

Cancer Data Integration and Querying with GeneTegra

E. Patrick Shironoshita[1], Yves R. Jean-Mary[1], Ray M. Bradley[1],
Patricia Buendia[1], and Mansur R. Kabuka[1,2]

[1] INFOTECH Soft, Inc. 1201 Brickell Avenue, Suite 220, Miami, FL 33131, USA
[2] University of Miami, Coral Gables, FL 33124, USA
{patrick,reggie,rbradley,paty,kabuka}@infotechsoft.com

Abstract. We present the GeneTegra system, an ontology-based information integration environment. We show its ability to query multiple data sources, and we evaluate the relative performance of different data repositories. GeneTegra uses Semantic Web standards to resolve the semantic and syntactic diversity of the large and increasingly complex body of publicly available data. GeneTegra contains mechanisms to create ontology models of data sources using the OWL 2 Web Ontology Language, and to define, plan, and execute queries against these models using the SPARQL query language. Data source formats supported include relational databases and XML and RDF data sources. Experimental results have been obtained to show that GeneTegra obtains equivalent results from different data repositories containing the same data, illustrating the ability of the methods proposed in querying heterogeneous sources using the same modeling paradigm.

Keywords: Data integration, ontology, Semantic Web, SPARQL, OWL.

1 Introduction

Outstanding achievements by the scientific community, such as the completion by the Human Genome Project of a high-quality, comprehensive sequencing of the human genome, have ushered in a new era in biological and biomedical research [1]. Nowadays there is a large, ever-growing, and complex body of bioinformatics and genetic data publicly available through the World Wide Web: the NAR Molecular Biology Database Collection lists over 1300 sources of data in its 2011 update [2]. This wealth of information is quite varied in nature and objectives, and provides immense opportunities to biomedical researchers, while posing significant challenges in terms of storing, accessing, and analyzing these data sets [1].

The ability to seamlessly access and share large amounts of heterogeneous data is crucial for the advancement of biomedical research, and requires resolving the semantic complexity of the source data and of the knowledge necessary to link this data in meaningful ways [3]. It is increasingly obvious that efficient integration solutions capable of working with large amounts of data in multiple formats are necessary to realize the enormous possibilities opened by the explosive growth in available biological, genetic, and clinical data.

O. Bodenreider and B. Rance (Eds.): DILS 2012, LNBI 7348, pp. 56–70, 2012.
© Springer-Verlag Berlin Heidelberg 2012

The essential problem in data integration is not in how to store it or retrieve it, but in how best to distill insights and associate these interpretations with the data [4]. Information integration is still a very inflexible process, which requires substantial manual intervention [5]. Changes in structure of just one data source could force a complete integration redesign [6]. Semantic technologies such as ontologies offer the promise of increased automation and standardized formal representation [7]. Knowledge representation techniques that can explicitly describe the semantics of the data are needed to improve interoperability for biological data representation and management [8]. Biomedical ontologies are an important element to provide for semantic interoperability and information exchange in biomedicine, and are increasingly used for the integration of data [9].

In the general realm of biological and biomedical applications, a number of integration systems and tools have been developed in the last few years. Integration approaches used in existing systems can be broadly classified into two categories: data warehousing, which deals with the translation of data into a centralized repository, and data federation, which employs query translation to decompose a global query into local queries at each integrated data source [9]. Surveys of recent approaches to data integration using ontologies and Semantic Web technologies can be found in [9,10]. Two recently developed systems bear particular mention. SWObjects [11] is a

Table 1. Mechanisms for ontology model generation

	Relational dB	**XML with XML Schema**	**caBIG data services**
Classes	Tables	Complex types	UML classes and NCIt concepts
Subclass relationships	None, as relational databases do not provide semantics that allow the determination of hierarchical relationships.	Derived types: if derived by extension, the extended type is the superclass; if by restriction, the restricted type is the subclass.	UML classes to their subclasses. Primary NCIt concepts to UML classes
Object properties	Key relationships between tables	Parent-child relationships between elements (modeled as properties, not as subclass relationships). Key relationships between elements.	UML associations
Datatype properties	Relationship between tables and field values	XML attributes and element content; the latter is related to its containing element using the property rdf:hasValue.	UML attributes.

collection of tools that enable federated querying using the SPARQL query language for RDF and query rewriting to SQL, but does not provide a comprehensive graphical interface. IO Informatics´ Sentient Suite [12] is a comprehensive application that can import content from multiple data and image formats, and can submit queries to relational databases and public sources, but seems to require loading of data into "meta-objects" rather than direct retrieval of results. Neither of these two current systems seems to have the ability of creating integrated data models composed of multiple sources, defining queries against them, distributing queries into separate sub-queries against each underlying source, and retrieving joined results.

In this article, we describe the mechanisms used to integrate multiple heterogeneous data sources within the GeneTegra system. GeneTegra uses ontology models to represent data sources, utilizes ontology alignment to assist in their integration, and provides the ability to create and execute queries against integrated views. To illustrate the ability of GeneTegra to integrate multiple data modeling paradigms, and to evaluate their relative performance, we present the results of experiments done using the same underlying data as the Corvus project in [13] loaded into different data source representations.

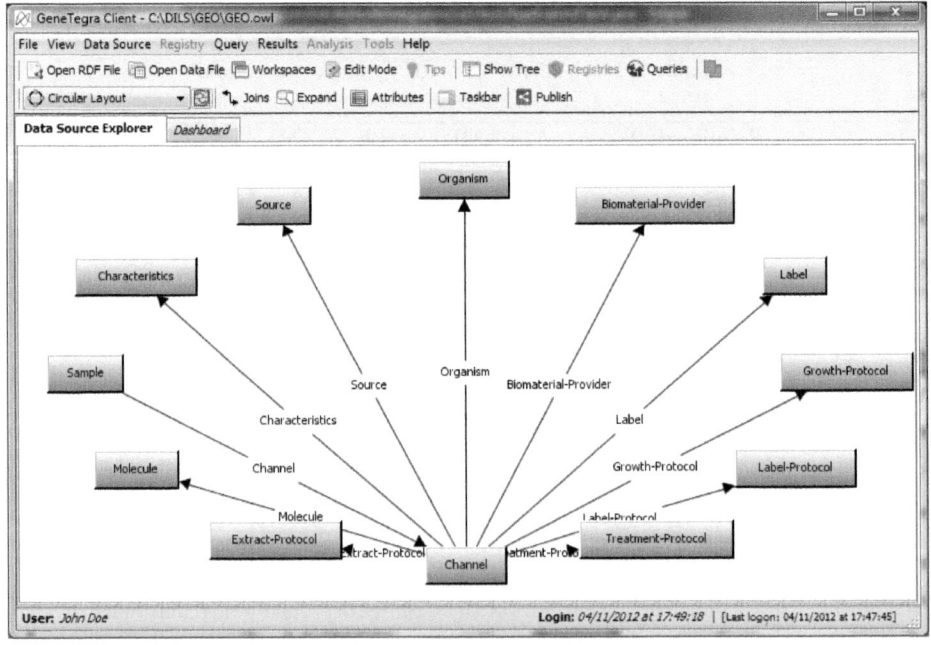

Fig. 1. Partial view of ontology model of Gene Expression Omnibus dataset. This partial view shows the relationships created for the Channel concept, including its incoming property relating it to a Sample, as well as outgoing relationships to different channel descriptors.

2   GeneTegra Information Integration

2.1   Ontology Representation of Data Sources

The GeneTegra information integration system contains mechanisms to represent the semantics of data sources using ontology models codified in the Web Ontology Language version 2 (OWL 2). It does so by first exploring the structure of data sources, and extracting the semantics of this structure. Methods have been developed to perform this process for relational databases, XML data sets, and RDF data stores, while taking advantage of the hierarchical organization already present in RDF data sets; in addition, the semCDI method to query caBIG® data services [14,15] has been incorporated in GeneTegra. A summary of the mechanisms for ontology model generation from different types of data sources is presented in Table 1. Each ontology model is assigned a unique namespace by GeneTegra, enabling the identification of classes and instances within the platform, while maintaining compatibility with Semantic Web standards.

Using a uniform modeling mechanism for all data sources enables the GeneTegra system to define queries that can be run against any underlying data representation. For example, consider a query as proposed in [13], where we wish to find the

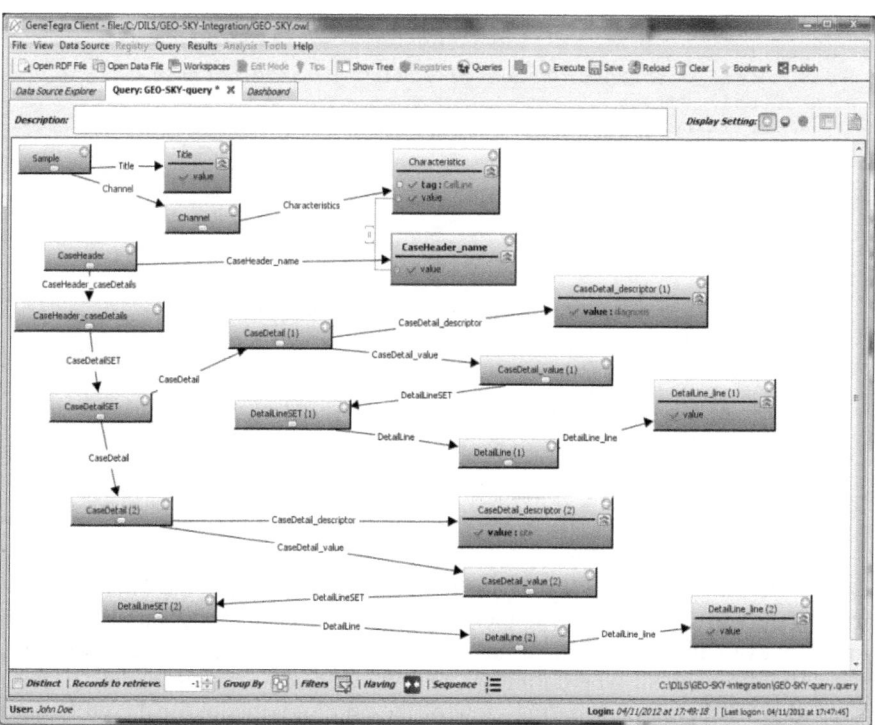

Fig. 2. Example query against SKY/M-FISH and GEO datasets. The query seeks to extract diagnoses in the SKY/M-FISH dataset that correspond to the samples in the GEO dataset.

Table 2. XML fragment depicting attribute-value pairs in GEO dataset. This table shows two XML fragments, from GEO and SKY/M-FISH respectively; (a) shows a GEO `Sample` node containing a set of `Characteristics` subnodes as attribute-value pairs, where the attribute name is given by the `tag` XML attribute. (b) shows a `CaseDetail` node with attribute names given by `CaseDetail_descriptor` and attribute values given four node levels down the `CaseDetail_value` node.

(a)

```
<Sample iid="GSM133995">
<Title>MCF7 hgu95a</Title>
<Accession database="GEO">GSM133995</Accession>
    ...
<Channel position="1">
<Organism taxid="9606">Homo sapiens</Organism>
<Characteristics tag="MDR Function">10</Characteristics>
<Characteristics tag="Prior Treatment">unknown</Characteristics>
<Characteristics tag="p53 Status">wild type</Characteristics>
<Characteristics tag="Age">69</Characteristics>
<Characteristics tag="BioSourceType">other[pleural effusion]
</Characteristics>
        <Characteristics tag="CellLine">MCF-7</Characteristics>
        ...
</Sample>
```

(b)

```
<CaseDetail>
   <CaseDetail_descriptor>casename</CaseDetail_descriptor>
   <CaseDetail_value>
      <DetailLineSET>
         <DetailLine>
            <DetailLine_line>HT-29</DetailLine_line>
         </DetailLine>
      </DetailLineSET>
   </CaseDetail_value>
</CaseDetail>
<CaseDetail>
   <CaseDetail_descriptor>diagnosis</CaseDetail_descriptor>
      <CaseDetail_value>
         <DetailLineSET>
            <DetailLine>
               <DetailLine_line>adenocarcinoma</DetailLine_line>
            </DetailLine>
         </DetailLineSET>
      </CaseDetail_value>
</CaseDetail>
```

diseases determined in the SKY/M-FISH data source [16] for the NCI-60 cell line samples listed in Gene Expression Omnibus (GEO) GEO-5949 family [17]. Both sources are available as XML datafiles, with the SKY/M-FISH data corresponding to the SLAII.dtd document type definition, and the GEO data corresponding to the MiNiML.xsd XML Schema. A partial view of the ontology model for the GEO dataset, showing the concepts related to the `Channel` object, can be seen in Figure 1. A similar model was created for the SKY/M-FISH dataset, and thus a query against an integrated view with both sources can be defined as shown in Figure 2.

2.2   Converting Attribute-Value Pairs to Concept-Instance Relationships

One paradigm in the creation of relational and XML data sources is the use of attribute-value pairs, to provide for flexibility in using different attributes while maintaining the same schema. For example, the MiNiML XML Schema on which the GEO data is based contains a node called `Characteristics`, which provides a set of characteristics for a sample. This node contains an attribute called `tag`, which defines a name of some characteristic, and a value; for example, Table 2(a) shows an XML fragment with a tag of `CellLine` and a value of `MCF7`. In other cases, the attribute and value may be even further apart, as the example in Table 2(b), taken from the SKY/M dataset, shows; it can be seen that the attribute name is separated from its corresponding value by four XML levels.

Attribute-value implementations are similar to RDF triples, in that they provide their semantics within the data itself, rather than as part of the metadata of the source. Assignment of semantic types in attribute-value pairs is done implicitly by the names of the attributes. In RDF, semantic type assignment can be more explicitly provided by linking an instance to an ontology class through an `rdf:type` relationship. GeneTegra is designed to make use of these relationships linking RDF instances to concepts encoded as ontology classes.

For attribute-value pairs in XML or relational databases, GeneTegra takes advantage of this functionality and of its ability to decouple the ontology model from the metadata for the source. In this manner, structures such as those found in GEO can be resolved by defining ontology classes based on the values of a given attribute. For example, the data from the example in Table 2(a) is mapped to an ontology model as illustrated in Figure 3, where the value for the tag attribute is used as the name for a concept, and the actual value of the element is modeled as an instance. A similar

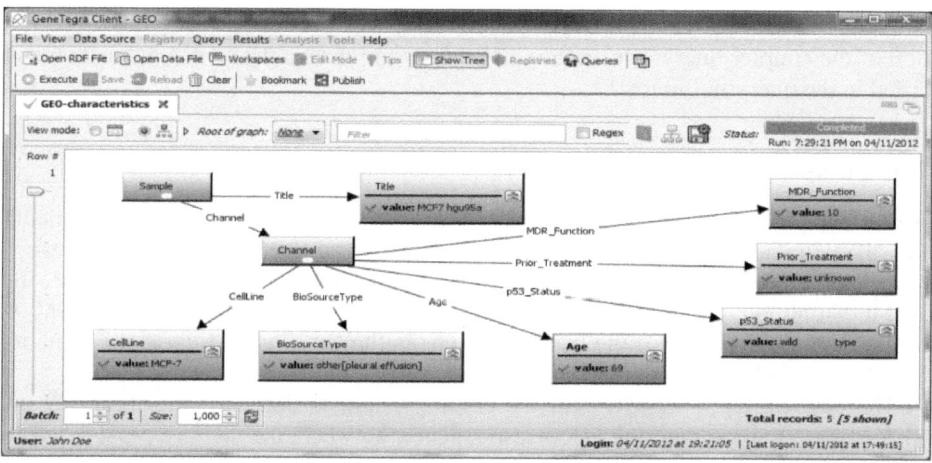

Fig. 3. Ontology model mapping from attribute-value pairs to concept-instance relationships. The characteristics of cell line MCF-7 shown in Table 2(a) are depicted graphically as modelled as concept-instance relationships mapped from the set of attribute-value pairs in the GEO `Characteristics` node.

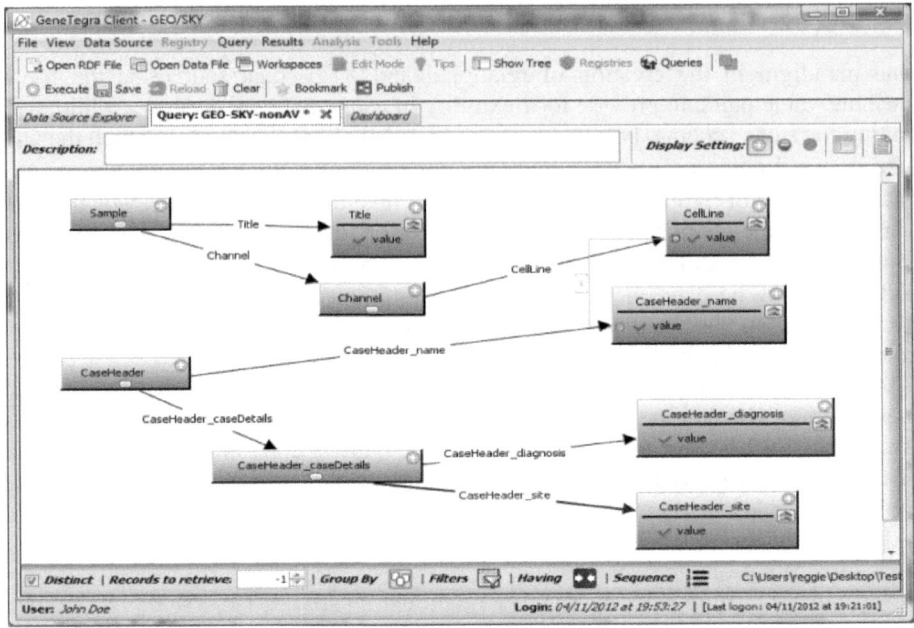

Fig. 4. Example query against SKY/M-FISH and GEO datasets with attribute-value pairs modeled as concept-instance relations. The query from Fig. 2 is made simpler through concept-instance modeling.

conversion can be done on the SKY/M dataset in Table 2(b), eliminating all intervening XML levels to create CaseDetail descriptors as objects with values given by the contents of the DetailLine_line node. These streamlined models then allow the creation of the simpler query, illustrated in Figure 4, to retrieve the same data as that specified in the query in Figure 2.

2.3 Linking Data Sources

An advantage of using ontology models for data sources is the ability to semantically link the extracted classes with each other, creating an integrated model from two or more data sources, independent of their underlying format. GeneTegra applies the ASMOV ontology alignment algorithm [18] to find such correspondences. ASMOV has been considered one of the best algorithms in the Ontology Alignment Evaluation Initiative (OAEI) contests [19].

The application of ASMOV to the ontology models derived from the GEO and SKY datasets does not find a correspondence between the Characteristics node in GEO and the CaseDetail node in SKY, since their lexical, structural, and extensional characteristics are all different. Applying the process of conversion from attribute-value pairs to concept-instance relationships, however, results in a set of concepts that can be better matched based on the actual data that they contain, such as

Table 3. SPARQL query from Figure 4 against GEO and SKY/M-FISH datasets with attribute-value pairs modelled as concept-instance relations

```
PREFIX   ns2:   <urn:its:xml:sky-concept/t-box#>
PREFIX   ns1:   <urn:its:xml:geo-concept/t-box#>
PREFIX   xsd:   <http://www.w3.org/2001/XMLSchema#>
PREFIX   rdf:   <http://www.w3.org/1999/02/22-rdf-syntax-ns#>
SELECT   *
WHERE
  {?c_Sample0 rdf:typens1:c_Sample ;
             ns1:c_Sample_Channel    ?c_Channel2 ;
             ns1:c_Sample_Title      ?c_Title1 .
   ?c_Channel2   rdf:typens1:c_Channel ;
             ns1:c_Channel_CellLine    ?c_CellLine3 .
   ?c_CellLine3   rdf:typens1:c_CellLine ;
rdf:value?c_CellLine3Value .
   ?c_CaseHeader_name5   rdf:typens2:c_CaseHeader_name ;
rdf:value?c_CaseHeader_name5Value .
   ?c_Title1    rdf:typens1:c_Title .
   ?c_CaseHeader4rdf:typens2:c_CaseHeader ;
             ns2:c_CaseHeader_caseDetails   ?c_CaseHeader_caseDetails6 ;
             ns2:c_CaseHeader_name ?c_CaseHeader_name5 .
   ?c_CaseHeader_caseDetails6
rdf:typens2:c_CaseHeader_caseDetails;
             ns2:c_CaseHeader_diagnosis   ?c_CaseHeader_diagnosis7 ;
             ns2:c_CaseHeader_site   ?c_CaseHeader_site8 .
   ?c_CaseHeader_diagnosis7
rdf:typens2:c_CaseHeader_diagnosis .
   ?c_CaseHeader_site8   rdf:type ns2:c_CaseHeader_site .
   OPTIONAL
{ ?c_Title1   rdf:value   ?c_Title1Value .}
   OPTIONAL
    { ?c_CaseHeader_diagnosis7
rdf:value   ?c_CaseHeader_diagnosis7Value .}
   OPTIONAL
     { ?c_CaseHeader_site8
rdf:value   ?c_CaseHeader_site8Value .}
     FILTER ( ?c_CellLine3Value = ?c_CaseHeader_name5Value )
  }
```

the CellLine concept on the GEO datasets and the CaseHeader_name on the SKY side, or Age on the GEO side and CaseHeader_age on the SKY side.

Such semantic correspondences provide hints to the users as to how the data sources could be joined together. In our case, we are interested in joining the datasets based on the link between CellLine and CaseHeader_name. GeneTegra allows the user to choose the actual concepts or relations on which joins should be done.

2.4   Query Distribution

Queries in GeneTegra are specified using the SPARQL Protocol and Query Language for RDF. These queries are created against the ontology models defined for data sources, or against integrated models obtained by semantically linking multiple sources together. Consider for example the query illustrated in Figure 4; its SPARQL

Table 4. Sub-query to run against GEO data

```
PREFIX   ns1:   <urn:its:xml:geo-concept/t-box#>
PREFIX   xsd:   <http://www.w3.org/2001/XMLSchema#>
PREFIX   rdf:   <http://www.w3.org/1999/02/22-rdf-syntax-ns#>
SELECT  *
WHERE {
{
        ?c_Channel2 rdf:type ns1:c_Channel
        ?c_Title1 rdf:type ns1:c_Title
        ?c_CellLine3 rdf:value ?c_CaseHeader_name5Value
        ?c_Sample0 ns1:c_Sample_Title ?c_Title1
        ?c_Sample0 rdf:type ns1:c_Sample
        ?c_Sample0 ns1:c_Sample_Channel ?c_Channel2
        ?c_CellLine3 rdf:type ns1:c_CellLine
        ?c_Channel2 ns1:c_Channel_CellLine ?c_CellLine3
}
OPTIONAL {
        ?c_Title1 rdf:value ?c_Title1Value
        }
}
```

representation is given in Table 3. Note that to obtain this SPARQL query, nodes referring to ontology classes have been mapped to variables linked to them through the `rdf:type` relationship; object properties linking classes have been used to relate these variables; and datatype properties, shown in the graphical layout as attributes of class nodes, have been mapped to variables expected to hold the data being queried.

If queries are defined against these integrated ontology models comprising multiple data sources, it is necessary to resolve the queries into sub-queries against each specific data source that potentially contains results pertaining to the query, and then perform inter-source operations to resolve the results into a single query result set. The process used by GeneTegra consists in the separation of the graph pattern of the query into multiple sub-graph patterns to be matched against different data sources.

First, every filter in the query is pushed down as far as possible into a query, to ensure that filters that reduce result sizes are executed early in the query plan. Then, each triple pattern in the query is examined to see whether it mentions any specific data source, as identified by the namespaces associated with the elements in the pattern. Note that GeneTegra keeps track of its generated namespaces, to distinguish them from namespaces that may exist either within the data sources themselves, such as references to FOAF or Dublin Core, or within the queries, such as the use of `rdf:type` to indicate instance membership in classes. All triple patterns that correspond to the same data source are grouped together, utilizing equivalences as defined in our semQA query algebra extension for SPARQL [20] to ensure that transformed queries return the same results as the originally posed queries. Specifically, the associative property of the `Join` operator and the left-associative properties of the `LeftJoin` operator in SPARQL allow for the reordering of triple patterns. As a result, the query in Table 3 is transformed into two sub-queries to be executed against different sources, which are shown in Tables 4 and 5, and their results retrieved and joined together using GeneTegra's query planning and execution mechanisms detailed

Table 5. Sub-query to run against SKY/M data

```
PREFIX  ns2:   <urn:its:xml:sky-concept/t-box#>
PREFIX  xsd:   <http://www.w3.org/2001/XMLSchema#>
PREFIX  rdf:   <http://www.w3.org/1999/02/22-rdf-syntax-ns#>
SELECT *
WHERE {
{
        ?c_CaseHeader_name5 rdf:value ?c_CaseHeader_name5Value
        ?c_CaseHeader4 rdf:type ns2:c_CaseHeader
        ?c_CaseHeader4 ns2:c_CaseHeader_CaseHeader_name
?c_CaseHeader_name5
        ?c_CaseHeader_diagnosis7 rdf:type ns2:c_CaseHeader_diagnosis
        ?c_CaseHeader_caseDetails6 rdf:type
ns2:c_CaseHeader_caseDetails
        ?c_CaseHeader_site8 rdf:type ns2:c_CaseHeader_site
        ?c_CaseHeader4 ns2:c_caseDetails ?c_caseDetails6
        ?c_CaseHeader_name5 rdf:type, ns2:c_CaseHeader_name
        ?c_caseDetails6 ns2:c_CaseHeader_caseDetails_site
?c_CaseHeader_site8
        ?c_CaseHeader_caseDetails6
ns2:c_CaseHeader_caseDetails_diagnosis ?c_CaseHeader_diagnosis7
}
OPTIONAL {
        ?c_CaseHeader_diagnosis7 rdf:value
?c_CaseHeader_diagnosis7Value
        }
OPTIONAL {
        ?c_CaseHeader_site8 rdf:value ?c_CaseHeader_site8Value
        }
}
```

in the next section. This query manipulation results in GeneTegra being able to take advantage of optimization processes that exist in each underlying data source to provide results efficiently, rather than requiring the retrieval of data for every single separate triple pattern.

2.5   Query Planning and Execution

Once queries against each specific data source are specified, query planning is performed to select the order in which to execute these queries. In the query in Figure 4, for example, two sub-queries must be done: one against the SKY/M instance and another against the GEO instance.

To perform this query, the system chooses which of the two sub-queries to execute first, based on an estimation of the number of results to be retrieved by each. The subquery with the smallest number of results is retrieved first; in this case, it will be the data from GEO, which contains data for cell lines, while SKY/M contains data for hybridizations. Then, a semi-join is performed, where the results from the query to GEO are used as query parameters to retrieve results from SKY/M. Each sub-query graph pattern defined against a data source is then converted to the native format defined for the source. The case of RDF-based sources is the simplest: the graph pattern

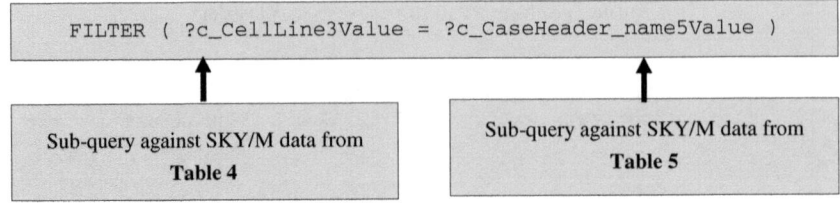

Fig. 5. Execution sequence of integrated query in Table 3

is simply wrapped with a SELECT query form and all results are retrieved as tables of variable bindings. In our current implementation, XML data sources are first transformed to RDF and then queried from RDF stores; it is planned that in the future an XPath/XQuery implementation will be available, to query directly from the XML dataset.

For relational databases, an SQL query is constructed from the corresponding graph pattern. To perform this transformation, we modified and extended the D2RQ implementation [21]: references from variables to OWL classes and properties are converted to SQL tables and fields, respectively, and filter conditions are established as part of the SQL WHERE clause. Unlike D2RQ, however, we identify object property relationships between classes as INNER JOIN clauses, and we also use a semi-join approach to resolve SPARQL OPTIONAL graph patterns, avoiding the execution of a full cross-product join and subsequent filtering.

caGrid data services use an XML-based query language specially designed for the environment, called caGrid Query Language (CQL). For these services, we create CQL queries in a similar fashion as with SQL, identifying UML classes and UML attributes from OWL classes and properties respectively. In this manner, a query is defined against instances of the caArray and caTissue caGrid data services, loaded with the SKY/M-FISH and the GEO data, respectively.

The execution of each sub-query results in a table of variable bindings to the retrieved data. Joining of the two tables is then done by processing the query elements that contain references to both sub-queries, in this case, a FILTER element, as shown in Fig. 5.

3   Results and Discussion

To demonstrate the versatility of GeneTegra in integrating multiple sources of different formats, the experiment described in [13] was executed against different configurations of the same data.

The two sets of data used were the manually curated data on the NCI-60 cell line obtained from SKY/M-FISH & CGH [16], and the data from the GEO experiment GSE5949 [17]. For querying purposes, this data was transformed into RDF and loaded into a Sesame RDF data store [22]. We labeled these datasets SKY-XML and GEO-XML respectively.

Since both datasets contain attribute-value pairs as detailed previously, we created a separate ontology model for each dataset with a transformation to a concept-instance representation; these models were labeled SKY-concept and GEO-concept.

In addition, we created locally a caGrid environment, and loaded the SKY/M-FISH &CGH and GEO data into caTissueSuite version 1.1 and caArray 2.3.1.1 instances respectively; each of these was labeled SKY-caGrid and GEO-caGrid. Moreover, to demonstrate the ability to work with relational databases, we created an ontology model of the SKY/M data only, by extracting a model from the MySQL database underlying the caTissue instance, and labeled this SKY-SQL. We should note that the clinical diagnoses in the SKY/M-FISH & CGH are less specific than those required by caTissueSuite. To perform this experiment, we used the same manual annotation performed by the authors of the Corvus project [13], with the exception of cell lines MDA-MB-231 and MDA-MB-435, which were annotated in [13] to a diagnosis of "Not specified". Given the diagnoses in SKY/M-FISH & CGH for these cell lines ("Adenocarcinoma of breast" and "Ductal carcinoma of breast" respectively), we annotated these two particular lines to appropriate diagnoses in caTissue ("Carcinoma of breast (disorder)" and "Intraductal carcinoma in situ of breast (disorder)" respectively).

All caGrid data services and RDF data sources were created running under VM-Ware virtual machines in a 2x quad-core Intel Xeon 1.86GHz computer with 8GB RAM; the Sesame RDF data store and caArray data service were configured to use 2 CPUs and 2 GB RAM running FreeBSD 8.0, while caTissue used 2 CPUs and 512MB RAM, also under FreeBSD 8.0. The GeneTegra query engine was run on a 2.26 GHz Intel Core 2 Duo computer with 4 GB RAM running Mac OSX 10.6.4. A 100-BaseT network connection was used between the query engine and the data sources.

We created SPARQL queries against combinations of these different models that had the SKY/M-FISH data on one side and the GEO data on the other, to obtain links between hybridizations and diagnoses. A comparison of the results obtained querying caTissue data with those done by the Corvus project as presented in [13] is shown in Table 6, in order to validate our results against this independent previous experiment. As can be seen, while most hybridizations are found by both systems, a few hybridizations were found by one method and not by the other. Investigation within the actual data loaded into the datasets used by our queries shows that the 5 hybridizations found by Corvus but not by GeneTegra do not exist within the caTissue data that was

Table 6. Comparison of query results between GeneTegra and Corvus. The table shows the total number of hybridization-diagnosis pairs found by both systems, and the number of differences between the systems.

	GeneTegra	Corvus
Total hybridizations found	295	285
Hybridizations found only by GeneTegra	15	
Hybridizations found only by Corvus		5

Table 7. Execution times for queries against different data models. The table shows the execution times for the queries ran against the different data and ontology models, for the SKY/M-FISH and the GEO data.

GEO	SKY/M-FISH	Execution Time (s)
GEO-XML	SKY-XML	16.55
GEO-XML	SKY-concept	15.70
GEO-XML	SKY-SQL	15.64
GEO-XML	SKY-caGrid	17.50
GEO-concept	SKY-concept	9.79
GEO-concept	SKY-SQL	9.95
GEO-concept	SKY-caGrid	11.65
GEO-caGrid	SKY-XML	191.22
GEO-caGrid	SKY-concept	190.89
GEO-caGrid	SKY-SQL	218.02
GEO-caGrid	SKY-caGrid	198.33

used in our experiments; they may have been removed in the time intervening between the two experiments. Similarly, we see that all 15 hybridizations that were found by GeneTegra but not found by Corvus exist within the datasets, so again, they could have been added in the intervening time.

The execution time of the different queries run is shown in Table 7. The execution times when running against the GEO XML data sets are an order of magnitude better than those obtained when running against the caArray caGrid data service. Execution times when running against the SKY/M-FISH & CGH XML data sets are also better than those obtained when running against caTissue, but they are not as significant. This degradation of performance in using caGrid can be mostly traced to the limitations in the CQL language used in caBIG, specifically, the "n+1 select" problem observed in [13]: in CQL, it is not possible to specify that two classes must be related by an attribute link, thus requiring that a set of values be retrieved from one object, and one query per value be created against the other object, and resulting in the derivation of multiple CQL queries for a single SPARQL query against the ontology model.

4   Conclusions

As has been shown, GeneTegra retrieves the same results when querying the source data from GEO and from SKY/M-FISH & CGH data using ontology models directly derived from the metadata, or using ontology models that seek to more closely depict the semantic characteristics of the data by converting attribute-value pairs to class-instance relationships. It also obtains the same results after loading the source data into instances of the grid services for caArray and caTissue, and when loading the SKY/M-FISH & CGH data into a relational database implementation of caTissue. The diversity of sources used provides evidence of the flexibility afforded by the use of OWL to model data sources, and by the GeneTegra mechanisms for querying multiple heterogeneous data sources through a single user interface and query language.

Our experiments show that the queries against XML data translated to RDF and loaded into Sesame were generally faster than those that required going through the caGrid data services; this is mainly because of the "n+1 select" problem with the CQL query language, but also attributable to overhead introduced by the caGrid environment in transforming queries from CQL into the Hibernate object-to-relational mapper and from there to SQL, and subsequently transforming results back from SQL to objects defined by Hibernate and then to the XML-based CQL query results format.

The experimental results also show that the modeling of attribute-value pairs as concept-instance relations produced a significant improvement, especially when done over the GEO data; this seems to be due to the lesser processing required in distributing the resulting simpler queries among the sources.

Our research team is continuing development of GeneTegra. We are currently working towards a tighter integration with OWL-based ontologies and ontology-defined data sources, In addition to the advantage of using controlled vocabularies, tighter links with ontologies is expected to potentiate the use of reasoning and inference, for example by retrieving data instances based upon inferred classifications over ontology classes. We are also developing a methodology for the execution of parameterized queries, where researchers will be able to access predefined, typical queries, alter certain parameters, and then access data without the need to construct an elaborate SPARQL query. Further, we are improving the D2RQ-based mechanisms for conversion from SPARQL to SQL to comply with the emerging R2RML recommendation from W3C [23]. Finally, we are looking at integration of data together with specific applications in genetics and biomedical research.

In this paper, we have shown the GeneTegra mechanisms for integrating data from relational databases, XML and RDF datasets, and caGrid data services. Thanks to the flexibility of the OWL ontology language and the power of the SPARQL query language for RDF, GeneTegra is able to create efficient models of these different data sources, and to obtain equivalent results regardless of the underlying data format.

Acknowledgements. This work is supported by grants # R44RR018667 from the National Center for Research Resources (NCRR), and grant # R44CA132293 from the National Cancer Institute (NCI). Both institutes are part of the U.S. National Institutes of Health (NIH).

References

1. Collins, F.S., Green, E.D., Guttmacher, A.E., Guyer, M.S.: A vision for the future of genomics research. Nature 422(6934), 835–847 (2003)
2. Galperin, M.Y., Cochrane, G.R.: The 2011 Nucleic Acids Research Database Issue and the online Molecular Biology Database Collection. Nucleic Acids Research 39(Database), D1–D6 (2010)
3. Brazhnik, O., Jones, J.F.: Anatomy of Data Integration. J. Biomed. Inform. 40(3), 252–269 (2007)
4. Neumann, E.: A Life Science Semantic Web: Are We There Yet? Sci. STKE 2005(283), pe22 (2005)

5. Bernstein, P.A., Haas, L.M.: Information integration in the enterprise. Commun. ACM 51, 72–79 (2008)
6. Roddick, J.F., de Vries, D.: Reduce, Reuse, Recycle: Practical Approaches to Schema Integration, Evolution and Versioning. In: Roddick, J.F., Benjamins, V.R., Si-said Cherfi, S., Chiang, R., Claramunt, C., Elmasri, R.A., Grandi, F., Han, H., Hepp, M., Lytras, M.D., Mišić, V.B., Poels, G., Song, I.-Y., Trujillo, J., Vangenot, C. (eds.) ER Workshops 2006. LNCS, vol. 4231, pp. 209–216. Springer, Heidelberg (2006)
7. Hepp, M., Leenheer, P.D., de Moor, A.: Ontology management: semantic web, semantic web services, and business applications. Springer (2007)
8. Dillon, T., Chang, E., Hadzic, M.: Ontology Support for Biomedical Information Resources. In: 21st IEEE International Symposium on Computer-Based Medical Systems, CBMS 2008, pp. 7–16 (2008)
9. Bodenreider, O.: Biomedical ontologies in action: role in knowledge management, data integration and decision support. Yearb. Med. Inform., 67–79 (2008)
10. Goble, C., Stevens, R.: State of the nation in data integration for bioinformatics. J. Biomed. Inform. 41(5), 687–693 (2008)
11. Prud'hommeaux, E., Deus, H., Marshall, M.S.: Tutorial: Query Federation with SWObjects. Nature Precedings,
http://precedings.nature.com/documents/5538/version/1
12. IO Informatics. Sentient Products Overview,
http://www.io-informatics.com/products/index.html
13. McCusker, J.P., Phillips, J.A., González Beltrán, A., Finkelstein, A., Krauthammer, M.: Semantic web data warehousing for caGrid. BMC Bioinformatics (10 suppl. 10),S2 (2009)
14. Shironoshita, E.P., Jean-Mary, Y.R., Bradley, R.M., Kabuka, M.R.: semCDI: a query formulation for semantic data integration in caBIG. J. Am. Med. Inform. Assoc. 15(4), 559–568 (2008)
15. Shironoshita, E.P., Bradley, R.M., Jean-Mary, Y.R., Taylor, T.J., Ryan, M.T., Kabuka, M.R.: Semantic Representation and Querying of caBIG Data Services. In: Bairoch, A., Cohen-Boulakia, S., Froidevaux, C. (eds.) DILS 2008. LNCS (LNBI), vol. 5109, pp. 108–115. Springer, Heidelberg (2008)
16. Knutsen, T., Gobu, V., Knaus, R., Padilla-Nash, H., Augustus, M., Strausberg, R.L., et al.: The interactive online SKY/M-FISH & CGH database and the Entrez cancer chromosomes search database: linkage of chromosomal aberrations with the genome sequence. Genes Chromosomes Cancer 44(1), 52–64 (2005)
17. Reinhold, W.C., Reimers, M.A., Lorenzi, P., Ho, J., Shankavaram, U.T., Ziegler, M.S., et al.: Multifactorial regulation of E-cadherin expression: an integrative study. Mol. Cancer Ther. 9(1), 1–16 (2010)
18. Jean-Mary, Y.R., Shironoshita, E.P., Kabuka, M.R.: Ontology matching with semantic verification. Web Semantics: Science, Services and Agents on the World Wide Web 7(3), 235–251 (2009)
19. Jean-Mary, Y.R., Shironoshita, E.P., Kabuka, M.R.: ASMOV: Results for OAEI 2010. In: Ontology Matching Workshop OM 2010 (2010)
20. Shironoshita, E.P., Jean-Mary, Y.R., Bradley, R.M., Kabuka, M.R.: semQA: SPARQL with Idempotent Disjunction. IEEE Transactions on Knowledge and Data Engineering 21(3), 401–414 (2009)
21. Bizer, C.: D2RQ - treating non-RDF databases as virtual RDF graphs. In: Proceedings off the 3rd International Semantic Web Conference (ISWC 2004) (2004), http://citeseer.ist.psu.edu/viewdoc/summary?doi=10.1.1.126.2314
22. Broekstra, J., Kampman, A., Van Harmelen, F.: Sesame: A Generic Architecture for Storing and Querying RDF and RDF Schema, pp. 54–68 (2002)
23. Das, S., Sundara, S., Cyganiak, R.: R2RML: RDB to RDF Mapping Language. W3C Candidate Recommendation, February 23 (2012), http://www.w3.org/TR/r2rml/

Integrating Large, Disparate Biomedical Ontologies to Boost Organ Development Network Connectivity

Chimezie Ogbuji and Rong Xu

Division of Medical Informatics, Center for Clinical Investigation
Case Western Reserve University, Cleveland OH 44106

Abstract. There is a significant opportunity to extend the Gene Ontology's (GO) anatomical development hierarchies for use in various bioinformatics analysis scenarios such as systems biology network analysis, for example. In particular there is very little overlap between the anatomical entities referenced in these hierarchies and the corresponding concepts in the Foundational Model of Anatomy (FMA) ontology, despite the fact that the FMA is quite vast and well organized. Both ontologies can be integrated in such a manner that new, biologically meaningful paths can be revealed for use in analysis. In this paper, we present a method for integrating the GO's anatomical development process hierarchies with anatomical concepts in the FMA that correspond to the organs participating in the development processes. We also provide an evaluation of the impact of this integration on the number of paths from diseases and disease genes to the FMA concepts that participate in the development processes that annotate the genes. The results demonstrate a vast number of such paths and therefore the potential to impact biological network analysis of the molecular mechanisms underlying diseases that are involved in anatomical development.

1 Introduction and Background

There is a significant amount of research into methods for use in analyzing the molecular basis of genetic diseases. Controlled biomedical vocabulary systems and ontologies play a major role in this. As a result, integrating these terminology systems, each of which may serve a different purpose, is an important biomedical informatics challenge. Their use often requires going from phenotypic phenomena to genetic phenomena and back. Furthermore, despite the abundance of these vocabularies and ontologies, there are many situations where portions are putatively known to be biologically related, but, their corresponding terminology cannot be readily used together.

In particular, in the case of two major reference biomedical ontologies that we will introduce below, there are no immediately usable terminology paths from a genetic disease to an anatomical entity that participates in the biological development processes used to annotate its associated genes. As an example, consider Marfan's Syndrome (MFS). MFS is defined [3] as

O. Bodenreider and B. Rance (Eds.): DILS 2012, LNBI 7348, pp. 71–82, 2012.

[..] mainly characterized by aneurysm formation in the proximal ascending aorta, leading to aortic dissection or rupture at a young age when left untreated. The identification of the underlying genetic cause of MFS, namely mutations in the fibrillin-1 gene (FBN1), has further enhanced our insights into the complex pathophysiology of aneurysm formation.

In the set of associations that are part of the Gene Ontology (GO), *Fibrillin 1* (FBN1) is annotated with two organ development biological processes: skeletal system development (**GO_0001501**) and heart development (**GO_0007507**). This association with the heart, for example, is not directly captured in biomedical ontologies despite the fact that - in this case - MFS's association with cardiovascular diseases such as aortic root dilation is well-documented in the literature [13,3,5]. Furthermore, recent work has discussed the new emphasis on aortic root aneurysm/dissection as a primary, diagnostic criterion for MFS [5]. Despite this fact, there are few if any direct terminological paths between MFS and aortic root aneurysm within the sources we use here (as we demonstrate later).

The GO primarily serves as a resource for the annotation of genes and gene products with its classes in such a way that the meaning expressed in the source of an annotation (whether from unstructured literature or a structured source such as a database) is captured in canonical syntax so that machines can interpret the underlying meaning interoperably. The GO has a hierarchy of biological processes involving organ development. In particular, this hierarchy is rooted at the *anatomical structure development* term (**GO_0048856**) which is defined as a "biological process whose specific outcome is the progression of an anatomical structure from an initial condition to its mature state."

Understanding the formation of the heart, one of the first organs formed as a result of an elegant coordination of integrated processes, is critical to the understanding of cardiovascular diseases (CVD) [9]. The study of genes and proteins involved in cardiovascular development is an important research area [16,10]. A recent effort was undertaken to expand the representation of GO biological processes involved in heart development and the result was a set of 281 new terms describing the anatomical representation of heart development and terms that describe processes that contribute to heart development: the Heart Development Gene Ontology (HDGO) [9].

In discussing various ways the GO and its artifacts can be used, the Gene Ontology Consortium's introduction to the Gene Ontology discusses how the existence of several ontologies allows the creation of 'cross-products' that maximize the utility of each ontology while avoiding redundancy. It goes further to say that by combining the developmental terms in the GO process ontology with a second ontology that describes Drosophila anatomical structures (for example), one could create an ontology of fly development.

Hoehndorf et al. provided [7] a method for making the intended meaning of phenotypic descriptions explicit and interoperable with anatomy and physiology ontologies. In describing the motivation for their work, they discuss how, in order to describe human developmental defects, a human developmental anatomy ontology integrated with the FMA would be beneficial as reference model on which deviations can be based.

The work described here seeks to facilitate that same approach for a handful of major anatomical structure development processes, their corresponding anatomical concepts, and the sub-categories and component-parts thereof, using the Foundational Model of Anatomy (FMA) as a basis.

The is_a relationship in both the GO and the FMA have the same meaning and are part of the common, underlying ontology format they both use to capture the meaning of their respective domains: the Web Ontology Language (OWL). The part_of and has_part relationships used by both have the same meaning and are part of the Relations Ontology (RO) [15] .

Most cellular components function via interaction with each other [1]. The complexity of such a resulting network can be quite formidable. As a result, understanding the context of a gene's network is an essential component of determining the phenotypic effect caused by disruptions or defects in such networks. Contemporary use of network and graph theory to analyze biological networks (disease networks, gene interaction networks, protein interaction networks, etc.) consistently reveals the importance of the characteristics of locally-dense and well-connected subgraphs (hubs) and network centrality measures [6,4,1,12,17]. Such subgraphs tend to indicate a *functional module*[1], and revealing additional, biologically-meaningful paths within them can be of great benefit.

We hypothesize that a high-quality integration of the GO's development process hierarchy with the FMA will have the following benefits (amongst others). First, it will provide new biological paths *from* genetic diseases *to* the organs whose development is involved in their molecular mechanisms. Secondly, it will provide biological paths in the reverse direction to facilitate the scoping of an analysis or similarity score to within the domain of a particular organ or bodily system. Finally, it will provide additional connectivity in biological networks to facilitate the use of graph and network analysis on them to find biologically-meaningful motifs.

By mapping *heart development* to *Heart* and *aorta development* to *Aorta*, there are now additional, biologically-meaningful paths between MFS and Aortic aneurysm beyond the direct path via FBN1. By adding such connections, one may find that these serve as an anatomical elucidation of relevant, immediate connections in a human disease network [6]. In figure 1, the *mutations-cause* connections between disorders and the genes whose mutations cause them are from the OMIM morbidity map. The *functionally_related_to* connections are from the GO annotations. The *has_part* and *is_a* connections between GO concepts and FMA classes are from the respective ontologies.

Intuitively, on the one hand, the aorta and the aortic valve are regionally-related, the latter participates in aortic valve development processes; all such processes are part of some heart development process, and FBN1 (which is causally associated with MFS) is putatively associated with heart development processes.

---

[1] An aggregation of nodes of similar or related function in the same network neighborhood [17].

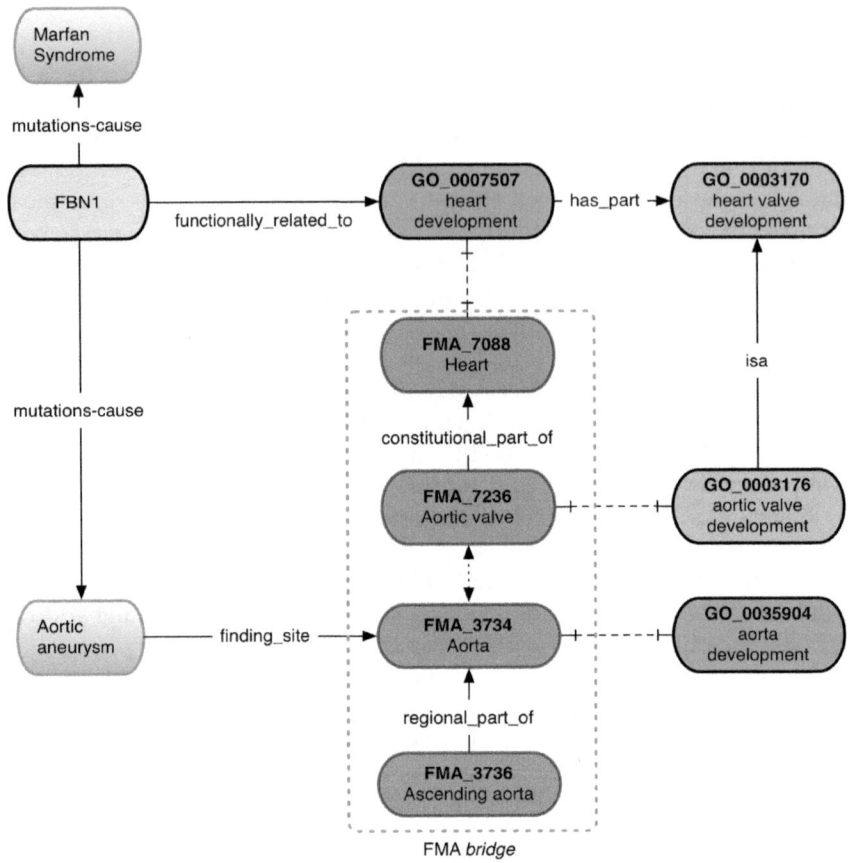

Fig. 1. Additional connectivity in MFS subnetwork

On the other hand, all aortas are components [14] of the heart, all hearts participate in heart development processes, and these are related to FBN1 as described previously[2]. In the diagram, the dashed box delineates the FMA-based bridge that facilitates additional connectivity between MFS and aortic aneurysm.

In this paper, we present a method for integrating the GO's anatomical development process hierarchies with anatomical concepts in the FMA that correspond to the organs participating in the development process. It attempts to replicate earlier successes in mapping the FMA to SNOMED CT [2] and subsequent efforts to leverage such a mapping to further integrate both ontologies in a manner that preserves the underlying meaning [11].

We provide results of measuring the overlap between hierarchies in both ontologies. We do this for various, major organ systems and their related processes for use as a basis for determining whether this approach is appropriate for a

---

[2] Paths of these kinds that go through GO development processes used to annotate human genes are presumably more meaningful for human diseases.

domain. Finally, we measure the amount of additional paths we now have as a result of the mapping.

2 Data and Method

In order to address many of the challenges described previously, the approach we use here takes advantage of several aspects of the representational technologies used with the ontologies involved. Both the GO and FMA are distributed in OWL and can therefore be hosted as a (named) RDF graph within a distributed RDF dataset available for query over the SPARQL RDF protocol. We stored the version of the GO made available on February 17th of 2012 and of the Open Biomedical Ontology (OBO) Foundry's[3] translation of the FMA that omits all relationships other than is_a, part_of, and has_part in a Virtuoso[4] server. They were queried from a process that began with the three root GO concepts of the following hierarchies:

- *Anatomical structure development*: the biological process whose specific outcome is the progression of an anatomical structure from an initial condition to its mature state.
- *Anatomical structure arrangement*: the process that gives rise to the configuration of the constituent parts of an anatomical structure.
- *Anatomical structure morphogenesis*: the process in which anatomical structures are generated and organized.

In each case, the algorithm iterated over all GO concepts subsumed by the root in an exhaustive manner (i.e., via consideration of the transitive properties of the is_a relationship). Only GO concepts that annotate human genes were considered. GO annotations of genes are distributed in subsets that are *taxon-specific* and the subset considered for use with this algorithm was the subset of the UniProt GO Annotations specific to Homo sapiens and available at the time of the version of the GO we used.

For each concept, depending on the hierarchy, the part of the label preceding the suffixes *development, structural organization, and morphogenesis* (respectively) was stemmed using the Natural Language ToolKit Python (NLTK) library's implementation of the Porter algorithm. This stem was used as the basis for dispatching a SPARQL query that returns FMA classes by lexical matching against the primary label or exact synonyms given, using a regular expression based on the stemmed words in the prefix.

Interacting with the biomedical ontologies in this way encourages a level of independence in the management of the ontology artifacts for distributed re-use from an algorithm such as ours which works directly off the ontology artifacts as input. In addition, this algorithm can exist as one of many analytical components that process biomedical OWL ontologies in this decoupled way.

---

[3] http://www.obofoundry.org/
[4] http://virtuoso.openlinksw.com/

The label of every FMA concept that matches this pattern (the primary label or an exact synonym) was compared in a case-insensitive manner with the prefix of the GO process concept to determine the *edit distance* ignoring leading or trailing white spaces. If either of the labels is an exact match for the prefix, this is considered a match.

3   Evaluation

We ran the mapping on the entirety of the 3 hierarchies. The resulting map is comprised of 1644 GO development processes and FMA anatomical entity pairs. However, the mapping algorithm we describe here is only as valuable as the cross-products that can be produced from it as well as the additional biological network paths that arise from it.

In order to quantify the potential value of the mapping, we calculated the Jaccard coefficient of the overlap between mapped hierarchies and measured the amount of additional paths introduced by the mapping. First, we started with a set of major organs (whose development processes are listed in the first column of the table below) and our mapping from a subset of the GO anatomical development processes (GO_{dev}) to a subset of FMA anatomical entities ($FMA_{concepts}$) that participate in them. GO_{dev} was limited to those development processes for which a mapping to a corresponding FMA anatomical entity was found.

For each root organ $\mathcal{O}$, we (exhaustively) calculated $GO_{subgraph(\mathcal{O})}$, which is comprised of all the GO concepts that are subsumed by $\mathcal{O}$ or are a component thereof.

Our running example demonstrates the interplay between the part-whole and specialization (is_a) relations. The heart development process GO concept (**GO_0007507**) has heart valve development as a component, which itself subsumes aortic valve development and has heart valve morphogenesis as a part. All three heart valve-related processes will be in $GO_{subgraph(\mathbf{GO\_0007507})}$.

Similarly, we calculated the set $FMA_{subgraph(\mathcal{O})}$ comprised of the FMA anatomical entities that are subsumed by $\mathcal{O}$ or are a component thereof. Then for a given anatomical entity $\mathcal{O}$, we calculated the set of GO terms that are both in $GO_{subgraph(\mathcal{O})}$ and for which there is a matching anatomical entity in $FMA_{subgraph(\mathcal{O})}$ that participates in the anatomical structure development, arrangement, and morphogenesis process denoted by each GO term. We refer to this set as $CS(\mathcal{O})$.

We computed the Jaccard coefficient of $FMA_{subgraph(\mathcal{O})}$ and $GO_{subgraph(\mathcal{O})}$, where $CS(\mathcal{O})$ is considered the intersection of both sets. Intuitively, this represents an overlap of terms used that (in part) considers the structure of the development process hierarchy. This is due to the fact that the sets compared are comprised of terms related via is_a and part_of. The table below shows the resulting Jaccard coefficients.

GO process	$GO_{subgraph(\mathcal{O})}$	$FMA_{subgraph(\mathcal{O})}$	$CS(o)$	Jaccard coefficient
pancreas development	5	55	3	$1.00 \cdot 10^{-2}$
liver development	2	337	1	$1.48 \cdot 10^{-3}$
brain development	155	1114	46	$2.30 \cdot 10^{-4}$
heart development	108	1289	26	$1.70 \cdot 10^{-4}$
kidney development	195	87	7	$1.30 \cdot 10^{-4}$
eye development	36	646	3	$1.20 \cdot 10^{-4}$

In order to quantify the additional biological paths that resulted from this mapping, we calculated the number of new paths from OMIM diseases through the genes they are associated with to the additional anatomy concepts. The only genes considered in this way are those that are annotated with some process in GO_{dev}. We will refer to this set as $Genes_{dev}$ and to the number of new paths as $\mathcal{P}^{+}_{dgo}$. We calculated the number of new paths from $Genes_{dev}$ to the newly connected anatomy concepts and will refer to this as $\mathcal{P}^{+}_{go}$.

Intuitively, $\mathcal{P}^{+}_{dgo}$ measures the amount of additional anatomical entities that could be said to participate in the development processes associated with the genes that underly a genetic disease. On the other hand, $\mathcal{P}^{+}_{go}$ measures the additional anatomical entities that can be associated with genes annotated by a development process in which either a more general anatomical entity or one that the new anatomical entity is a component of participates. This measure can provide some sense of the size of a disease gene network that can be constructed using a topology that links genes that share a common anatomical entity that participates in the development processes that annotate them.

The values for $\mathcal{P}^{+}_{dgo}$ are shown below for 847 of the 13,756 OMIM diseases associated with genes annotated with our matching development processes, leaving out those diseases that are not associated with a gene in $Genes_{dev}$. On average, the mapping introduces 9,549 additional biological paths from each such OMIM disease to a participating anatomical entity. Figure 2a shows the number of additional paths in a logarithmic scale plot and figure 2b is a histogram of both the distribution of the values of $\mathcal{P}^{+}_{dgo}$ as a whole and normalized by the gene (i.e., divided by the number of genes associated with the OMIM disease).

As we can see, for a majority of the diseases that had additional paths, the number of paths was well above 100. However, towards the upper end of the scale, the total number of paths surpasses the number of paths per gene. This reflects the intuition that the strength of the association between each gene and the disease it is causally associated with via OMIM is (roughly) inversely proportional to the total number of genes associated with the same disease.

In figure 3 we give both the top 20 OMIM diseases, ordered by $\mathcal{P}^{+}_{dgo}$ as well as the number of genes associated with it (not all of which are in $Genes_{dev}$) and the top 20 ordered by new paths per gene.

The values for $\mathcal{P}^{+}_{go}$ for all disease genes g in $Genes_{dev}$ from the OMIM are also given in a logarithmically-scaled histogram plot in figure 4, for only those genes that saw an increase in paths to anatomical structures. On average, each such gene had 17,037 additional paths to FMA anatomical entities. In this plot, the number of new paths per gene is less pronounced than in the case for $\mathcal{P}^+_{dgo}$. This reflects the fact that paths from diseases to organs introduce the combinatorial factor of disease-gene pairings.

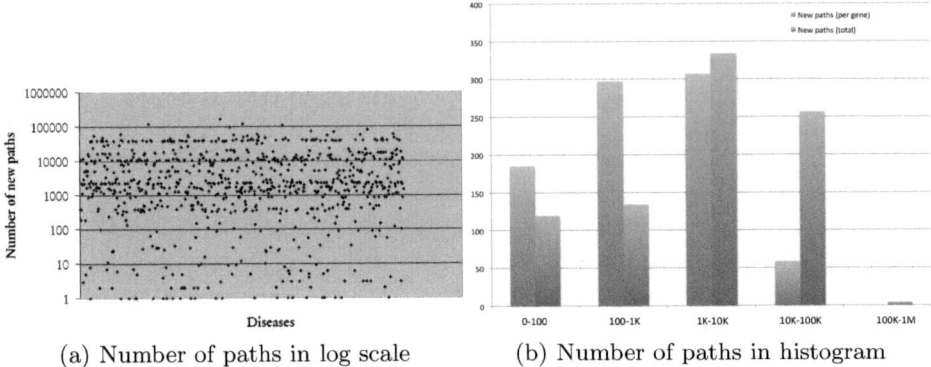

(a) Number of paths in log scale   (b) Number of paths in histogram

Fig. 2. $\mathcal{P}^+_{dgo}$

OMIM Disease	Number of genes	
Microphthalmia	48	164129
Epiphyseal dysplasia	21	116719
Pituitary hormone deficiency	12	116270
Mental retardation	210	104565
Leukemia	111	94665
Colorectal cancer	60	78687
Deafness	216	70573
Brachydactyly	22	67888
Chondrodysplasia punctata	9	59310
Coloboma	8	58965
Cardiomyopathy	145	57340
Myopathy	66	56003
Spondylocostal dysostosis	6	55765
Cataract	71	55512
Ehlers-Danlos syndrome	18	54255
Muscular dystrophy	69	54159
Melanoma	16	52007
Diabetes mellitus	38	50116
Jackson-Weiss syndrome	8	50089
Pfeiffer syndrome	8	50089

(a) Ordered by $\mathcal{P}^+_{dgo}$

Fig. 3. Top 20 OMIM diseases

OMIM Disease	genes	
Osteoporosis	VDR	40155.0
Rickets	VDR	40155.0
Ulnar-mammary syndrome	TBX3	39941.0
Intrauterine and postnatal growth retardation	IGF2	39101.0
Arrhythmogenic right ventricular dysplasia 1	TGFB3	38617.0
Lacticacidemia due to PDX1 deficiency	PDX1	38318.0
Insulin-like growth factor I	IGF1R	38316.0
Costello syndrome	HRAS	38305.0
GAMT deficiency	GAMT	38305.0
Supravalvar aortic stenosis	ELN	38305.0
Guttmacher syndrome	HOXA13,HOX1J	21094.5
Hand-foot-uterus syndrome	HOXA13,HOX1J	21094.5
Segawa syndrome	TYH,TH	19311.5
Lymphoplasmacytoid lymphoma	PAX5,BSAP	19157.0
Alagille syndrome 2	AGS2,NOTCH2	19157.0
Retinitis pigmentosa-17	RP17,CA4	19152.5
T-cell immunodeficiency	WHN,FOXN1	19152.5
DiGeorge/velocardiofacial syndrome complex-2	DGS2,DGCR2	19152.5
Growth retardation with deafness $\cdots$	IGF1	17857.0
Pilomatricoma	CTNNB1	17698.0

(b) Ordered by normalized $\mathcal{P}^{+}_{dgo}$

Fig. 3. *(Continued)*

4   Discussion

The first result from our evaluation, the Jaccard coefficients, indicates very little overlap between both the GO anatomical development hierarchies and the FMA hierarchies of anatomical entities that participate in them. This is not surprising as both ontologies cover very disparate domains within medicine and one is specific to humans whereas the other is species-agnostic. It is not clear from [9] whether or not the development process hierarchies in the GO, take heart development for example, uses a species-independent ontology of canonical anatomy as the basis for those development processes involving the formation of anatomy. In any case, the combination of the low overlap, the size of the FMA as a whole, and the size of the hierarchies matched by our process indicates there is an opportunity to build on this mapping and possibly integrate both ontologies in a meaningful way.

The second set of results demonstrate a significant increase of biological paths from both diseases whose molecular mechanisms are associated with the development of anatomical entities as well as the genes associated with them to the anatomy participating in the development process. Certainly, not all these paths will be of immediate value for analysis of the molecular mechanisms (related to the development of anatomy) underlying congenial disorders, for example. However, given that these paths are at least as logically and biologically sound as

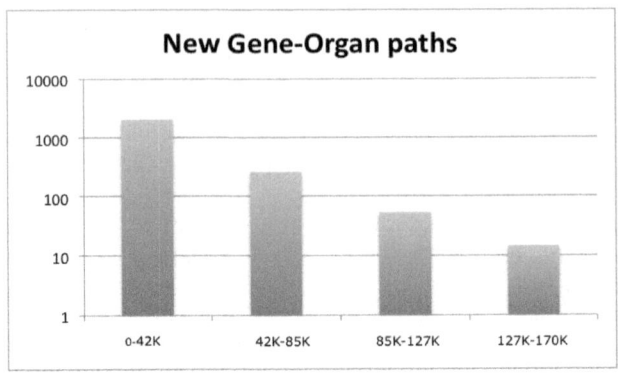

Fig. 4. Histogram of additional paths per $g \in Genes_{dev}$

the ontologies they were integrated from, we expect that an appreciable amount will be informative for analysis of the underlying biological networks.

To our knowledge, this is the first attempt of this kind to integrate the anatomical structural development, morphogenesis, and organization hierarchies in the GO with the FMA. The only mention we could find of logical connections between these two parts of each ontology is in the preliminary, logic-based evaluation of an integration of post-composed phenotype descriptions with domain ontologies (of which the FMA is one) by Jimenez-Ruiz et al. [8].

In that work, they describe unintended consequences due to the logical definitions restricting the range of the relations in the FMA and the use of the part_of relation between anatomical entities on the one hand and biological processes on the other. In what we describe here, since we query the ontologies separately in order to perform our analysis, we do not have to deal with this issue. In particular, our algorithm queries the GO and the FMA independently and in each case, the part_of relation is consistently used either between processes or continuants.

Another point of consideration is the fact that GO annotations are species-specific, whereas the FMA is a reference ontology for *human* anatomy. As described in [9], in relation to the relevance to gene and gene products involved in many congenital heart diseases, the GO is designed for use in annotating all gene products across all species in order to make (logical) conclusions about conserved or diverse biological processes. Logical entailments that follow from the integration of both the FMA and GO, for example, need to be careful to only consider annotations for humans or to have a robust way to manage the uncertainty introduced in not doing so.

5   Conclusion

We have presented a reproducible method for integrating the GO's anatomical development process hierarchies with anatomical concepts in the FMA that correspond to the organs participating in the development process. We have also provided a thorough evaluation of the impact of this integration in terms of the number of new paths from diseases and diseases genes to FMA concepts that participate in development processes. The results demonstrate a very large number of such paths and therefore potential to impact biological network analysis of the molecular mechanisms underlying congenital disorders as well as other diseases that are involved in anatomical development.

Our work can be further built upon by building gene, disease, and organ interaction networks on the basis of our mapping and the mapping can be improved via the use of the UMLS metathesaurus[5] names for the FMA as an additional lexicon to use in improving the ability to match alternative synonyms for an anatomy concept of interest. Altogether, we believe this serves as a strong foundation for studying the molecular mechanisms underlying congenital disorders and other diseases involving anatomical development.

References

1. Barabási, A.L., Gulbahce, N., Loscalzo, J.: Network medicine: a network-based approach to human disease. Nature Reviews Genetics 12(1), 56–68 (2011)
2. Bodenreider, O., Zhang, S.: Comparing the representation of anatomy in the FMA and SNOMED CT. In: AMIA Annual Symposium Proceedings 2006, p. 46. American Medical Informatics Association (2006)
3. De Backer, J., Loeys, B., De Paepe, A.: Marfan and Marfan-like syndromes. Artery Research 3(1), 9–16 (2009)
4. Emmert-Streib, F., Dehmer, M.: Networks for systems biology: Conceptual connection of data and function. IET Systems Biology 5(3), 185–207 (2011)
5. Faivre, L., Collod-Beroud, G., Adès, L., Arbustini, E., Child, A., Callewaert, B.L., Loeys, B., Binquet, C., Gautier, E., Mayer, K., et al.: The new ghent criteria for marfan syndrome: What do they change? Clinical Genetics (2011)
6. Goh, K.I., Cusick, M.E., Valle, D., Childs, B., Vidal, M., Barabási, A.L.: The human disease network. Proceedings of the National Academy of Sciences 104(21), 8685 (2007)
7. Hoehndorf, R., Oellrich, A., Rebholz-Schuhmann, D.: Interoperability between phenotype and anatomy ontologies. Bioinformatics 26(24), 3112 (2010)
8. Jimenez-Ruiz, E., Grau, B.C., Berlanga, R., Rebholz-Schuhmann, D.: First steps in the logic-based assessment of post-composed phenotypic descriptions. Arxiv preprint arXiv:1012.1659 (2010)
9. Khodiyar, V.K., Hill, D.P., Howe, D., Berardini, T.Z., Tweedie, S., Talmud, P.J., Breckenridge, R.: The representation of heart development in the gene ontology. Developmental Biology (2011)
10. Musunuru, K., Domian, I.J., Chien, K.R.: Stem cell models of cardiac development and disease. Annual Review of Cell and Developmental Biology 26, 667–687 (2010)
11. Ogbuji, C., Arabandi, S., Zhang, S., Zhang, G.Q.: Segmenting and merging domain-specific ontology modules for clinical informatics. In: Proceeding of the 2010 conference on Formal Ontology in Information Systems: Proceedings of the Sixth International Conference (FOIS 2010), pp. 414–427 (2010)

---

[5] http://www.nlm.nih.gov/research/umls/knowledge_sources/metathesaurus/index.html

12. Özgür, A., Vu, T., Erkan, G., Radev, D.R.: Identifying gene-disease associations using centrality on a literature mined gene-interaction network. Bioinformatics 24(13), i277–i285 (2008)
13. Roman, M.J., Rosen, S.E., Kramer-Fox, R., Devereux, R.B.: Prognostic significance of the pattern of aortic root dilation in the marfan syndrome. Journal of the American College of Cardiology 22(5), 1470–1476 (1993)
14. Rosse, C., Mejino, J.L.V.: The foundational model of anatomy ontology. Anatomy Ontologies for Bioinformatics, 59–117 (2008)
15. Smith, B., Ceusters, W., Klagges, B., Köhler, J., Kumar, A., Lomax, J., Mungall, C., Neuhaus, F., Rector, A.L., Rosse, C.: Relations in biomedical ontologies. Genome Biology 6(5), R46 (2005)
16. Sperling, S.R.: Systems biology approaches to heart development and congenital heart disease. Cardiovascular Research 91(2), 269 (2011)
17. Vidal, M., Cusick, M.E., Barabasi, A.L.: Interactome networks and human disease. Cell 144(6), 986–998 (2011)

Validating Candidate Gene-Mutation Relations in MEDLINE Abstracts via Crowdsourcing

John D. Burger[1], Emily Doughty[2], Sam Bayer[1], David Tresner-Kirsch[1],
Ben Wellner[1], John Aberdeen[1], Kyungjoon Lee[3],
Maricel G. Kann[2], and Lynette Hirschman[1]

[1] The MITRE Corporation, Bedford, MA, USA
{john,sam,davidtk,wellner,aberdeen,lynette}@mitre.org
[2] University of Maryland, Baltimore County, Baltimore MD, USA
{doughty2,mkann}@umbc.edu
[3] Harvard Medical School, Boston MA, USA
joon_lee@hms.harvard.edu

Abstract. We describe an experiment to elicit judgments on the validity of gene-mutation relations in MEDLINE abstracts via crowdsourcing. The biomedical literature contains rich information on such relations, but the correct pairings are difficult to extract automatically because a single abstract may mention multiple genes and mutations. We ran an experiment presenting candidate gene-mutation relations as Amazon Mechanical Turk *HITs* (human intelligence tasks). We extracted candidate mutations from a corpus of 250 MEDLINE abstracts using EMU combined with curated gene lists from NCBI. The resulting document-level annotations were projected into the abstract text to highlight mentions of genes and mutations for review. Reviewers returned results within 36 hours. Initial weighted results evaluated against a gold standard of expert curated gene-mutation relations achieved 85% accuracy, with the best reviewer achieving 91% accuracy. We expect performance to increase with further experimentation, providing a scalable approach for rapid manual curation of important biological relations.

Keywords: Crowdsourcing, mutations, text mining, annotation.

1   Introduction

The decrease in genome sequencing costs has led to an explosion of information on mutations in the human genome and their relation to disease. There are now a number of curated databases devoted to the capture of information on mutations in specific populations with associated phenotypes, including, e.g., OMIM [1], dbSNP [2], HGVbase [3], HGMD [4] and locus specific databases (LSDBs) [5]. Such databases are critical for downstream data integration, making it possible, for example, to interpret patient genetic data (SNPs in genes) against variants known to be associated with disease. These resources are also critical for genome-wide association studies (GWAS) [6], phenotype-wide association studies (PheWAS) [7] and pharmacogenomics [8]. However, much of the

O. Bodenreider and B. Rance (Eds.): DILS 2012, LNBI 7348, pp. 83–91, 2012.

information about these gene-mutation-disease associations is currently buried in the biomedical literature. There is increased demand to identify these results in a timely fashion and to make them available in a computable form to support personalized medicine and translational medicine applications.

In this paper, we focus on the curation challenges for a subportion of this problem: identifying relationships between genes and mutations from the biomedical literature. We investigate the efficacy of *crowdsourcing* to validate gene-mutation relations in MEDLINE abstracts. Crowdsourcing is the process of leveraging a large potential pool of non-experts for problem solving via the Web. Here we employ crowdsourcing to verify whether candidate gene-mutation associations in MEDLINE abstracts represent valid associations, using automated techniques to identify candidate genes and mutations in abstracts for presentation to reviewers.

Identifying relevant associations automatically in the literature requires two steps: identifying the elements (genes and mutations), and determining which ones are related. For MEDLINE abstracts, one of these sets of elements (genes) is routinely annotated by indexers at the National Library of Medicine, but mutation terms must be extracted directly from the abstract text, generally via regular expressions and/or natural language processing methods.

Mutations are challenging to identify in the literature because of the many types of mutations (SNPs, indels, rearrangements) and the variability in how the mutations are described in the literature. For example, SNPs (single nucleotide polymorphisms), the simplest class of mutation, can be described as a mutation in the DNA (in terms of nucleotides) or as a mutation in the protein (in terms of amino acids). Examples of SNPs found in the literature include condensed forms, such as "313A>G", "E161K", "Pro582Ser" or "AGG to AGT", as well as whole phrases, such as "substitution of Met-69 by Ala or Gly in TEM-1 beta-lactamase". In spite of these complexities, locating a subset of these mutations (especially SNPs) in the literature has been shown to be a tractable task for text-mining tools based on pattern matching, as shown by MutationFinder [9], Extractor of MUtations (EMU) [10], MutationTagger [11] and elsewhere [12–14].

The focus of the experiment described here is to establish the correct association between a mutation and the associated gene, otherwise known as *mutation grounding* [11]. When multiple mutations and genes are found in the same text, the association of mutations to genes is challenging, leading to false positives. Using EMU to extract mutations from abstracts, Doughty et al. [10] reported a high precision on detection of the mutation patterns (99 and 96% for prostate and breast cancer mutations, respectively) but a significant decrease in precision, of up to 20%, when attempting to automatically extract the correct gene-mutation pairs. Mutation grounding can be improved by filtering based on match of wild-type sequence to a reference sequence, given gene and positional information, as described in [10, 11]; however, this results in a decrease in recall. For these reasons, the mutation grounding task seemed like an interesting candidate for crowdsourcing.

2   Data

A gold standard gene-mutation-disease data set was created for three diseases (breast cancer, prostate cancer, and autism spectrum disorder) as follows. Mutation-related MEDLINE abstracts were downloaded from the PubMed search engine using the MeSH terms "Mutation" (breast and prostate cancer) or "Mutation AND Polymorphism/genetic" (autism spectrum disorder). Abstracts related to breast (5967 citations) and prostate cancer (1721 citations) were identified by MetaMap [15] as described in [10]. An additional set of abstracts related to autism spectrum disorder were identified using the MeSH term "Child Development Disorders, Pervasive." The EMU tool was used to identify mutations in the three disease abstract corpora. Abstracts for which EMU identified at least one mutation were selected for expert curation, resulting in a corpus of 810 MEDLINE abstracts, with 1573 mutations; almost 50% of the abstracts had two or more curated gene-mutation relationships.

For the initial crowdsourcing experiment, we focused only on the gene-mutation relations in the gold standard. Because of limitations of time and budget, we used a randomly selected 250-abstract subset of the full gold standard. Of these, six contained no mutations, 99 contained a single mutation, and the remaining 145 contained two or more mutations. The subset contained 568 gold-standard mutations altogether.

3   Methods

3.1   Overall Experimental Design

We designed the gene-mutation relation validation as a two-stage process. The first stage automatically identified all candidate mutations and genes in each abstract and then projected these candidates back into the text to highlight the specific mentions. The second stage utilized Amazon Mechanical Turk, an online crowdsourcing platform, to elicit judgments from multiple reviewers on correct gene-mutation pairings.

In stage 1, we used a modified version of EMU to extract and normalize mutations. For genes, we took the union of the NLM-curated genes associated with each abstract in PubMed and any additional genes extracted by EMU. We then generated all possible gene-mutation pairs for each abstract and constructed a separate item for each pair in each abstract for presentation to the reviewers. Each item is a yes/no question asking whether the abstract discusses the candidate relation between the highlighted gene and mutation. In stage 2, the items were distributed via Amazon Mechanical Turk, and the results were aggregated to provide judgments on the validity of each presented gene-mutation pair. To evaluate, we compared the aggregated reviewer judgments on gene-mutation pairs to the gold standard data. We also compared the output of the first stage pre-processing to the gold standard data, to determine loss of recall in preparing the data, and amount of over-generation.

3.2   Extraction of Candidate Genes, Mutations and Their Mentions

EMU [10] extracts point mutations from text and identifies relevant genes found in the input text. Gene identification is done using string matching against a customized list of gene names derived from the Human Genome Organization (HUGO) and NCBI Gene databases. To ensure maximal coverage of potentially relevant gene names, we augmented the genes identified by EMU with any additional genes curated for each article by NCBI. We then found all occurrences of these genes in the input text using exact string matching. In the case of NCBI genes, we searched for occurrences of the gene symbol, the gene name and any synonyms for the gene that appeared in the database. For mutations, EMU detects SNPs through a two-step filter. The first step collects likely mutation spans using a list of positive regular expressions; the second rejects candidates based on a stop-list of negative regular expressions. For both genes and mutations, we identified the text spans in the input text associated with each gene mention. To combine and visualize these tagged text spans, we relied on tools for human linguistic annotation found in the MITRE Identification Scrubber Toolkit [16]. These tools allowed us to create rich documents with *standoff annotations* that identify the type and location of the mention. Standoff annotations are not embedded in the text, and are thus amenable to manipulation and processing. The toolkit also provided a browser-based presentation library for highlighting the location of these mentions.

3.3   Amazon Mechanical Turk for Gene-Mutation Association Identification

Amazon Mechanical Turk (MTurk) is a web-based labor market for *Human Intelligence Tasks (HITs)*. HITs are typically minimal problems that cannot easily be automated, such as categorizing a photograph or a text snippet. Most HITs take a few seconds for a worker to perform, and pay a few cents. In 2011, Amazon indicated that there were over 500,000 workers (Turkers) registered, although as with all online services, many more people sign up than are active at any time. A number of researchers have recently experimented with the use of MTurk to create and annotate human language data [17]. In particular, MTurk has been used to annotate medical named entities in clinical trials descriptions [18].

To prepare candidate gene-mutation pairs for presentation to the Turkers, we first grouped together multiple mentions of the same gene, and also multiple mentions of each mutation. For genes, we used the gene ID, or if unavailable, the gene name. Mutations were canonicalized according to EMU by a triple of position, wild type and mutant nucleotide or amino acid. We took the cross product of all genes and all mutations found in an abstract, which resulted in 1299 pairs. Each of these gene-mutation candidates was presented as a separate HIT on MTurk. For each HIT, the Turker was asked a yes-or-no question to determine whether the given mutation and gene are in fact related. (Turkers were also allowed to choose *inconsistent annotation* to indicate a problem in the

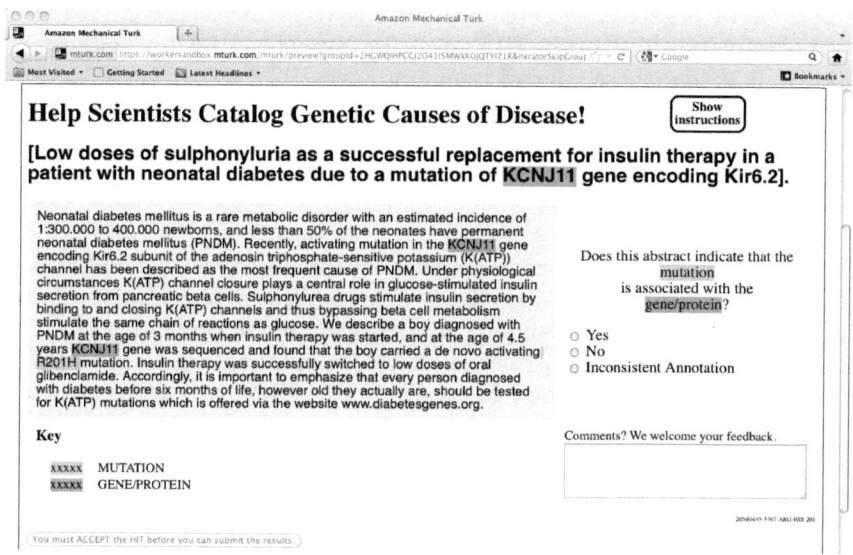

Fig. 1. HIT design for the gene-mutation task

projection–this option was rarely used, but was counted as a *no* answer.) The mentions of the gene and mutation in the relevant pair were highlighted in the abstracts text using the projection method described above. Figure 1 shows a screen capture of the HIT design, in which a single gene-mutation pair candidate is presented.

Because Turkers are typically not pre-qualified for any particular task, there are several strategies that requesters use to screen Turkers for tasks requiring specific expertise or skills. The requester can require that Turkers be from a particular region, or have a certain minimum approval rating on their previous HITs. In addition, the requester can leverage redundancy by asking for multiple Turkers to perform each HIT. The responses can then be aggregated in some way, perhaps by simply choosing the majority response. Another strategy is to insert *control items* into the HITs. These are items for which the correct answer is already known. Each Turker has a persistent ID that accompanies their responses, and by aggregating control item responses, each Turker's efficacy can be measured.

We used all of these strategies to control for Turker variability. First, we required the Turkers for these HITs to be from the United States, in order to minimize second language effects. We also limited participation to Turkers with 95% approval ratings, and we requested that five separate Turkers annotate each HIT. Finally, we included a number of control items. These were 44 gene-mutation pairs from a handful of separate abstracts developed in initial annotation experiments. The control items were identified as relatively easy to annotate–that is, even non-experts should agree as to whether a candidate pair was positive or negative. We created multiple HITs for each control item such that approximately 25% of the HITs each Turker encountered were these known items. By injecting these control items into the HITs, we are able to estimate the efficacy of the Turkers as described above.

4   Results

We measured how well the pre-processing captured the relevant mutations and genes by comparing them independently against the sets of mutations and genes that appeared in the gold-standard curated gene-mutation associations. In the 250 abstracts in our dataset, there were 568 unique mutations at the abstract level in the gold standard; the pre-processing via EMU correctly extracted 484 (recall of 85.2%). There were 279 unique genes at the abstract level; the gene pre-processing using EMU and NCBI gene lists extracted 260 of these genes (recall of 93.2%).

Overall, there were 1299 candidate gene-mutation pairs presented to the Turkers, and 568 unique mutation-gene pairs in the gold standard, giving a distribution of 44:56 positive to negative candidates. Of the 250 abstracts, there were 181 for which all of the mutations and genes that appeared in the curated gene-mutation pairs were identified by our automated methods. These abstracts are of particular interest for our experiments since all of the curated gene-mutation relations for these documents could, in theory, be identified through crowdsourcing. We refer to these below as *perfect documents*.

1733 HITs were posted to MTurk, including 1299 candidate gene-mutation pairs generated from the 250 abstracts, with the rest being control items as described above. We requested that five Turkers work on each HIT, for a total of 8665 judgments. These were completed within 36 hours of posting them. Altogether, 23 Turkers responded, although seven only performed one HIT. Eight Turkers performed ten or more HITs, and six performed 100 or more. We paid 7¢ per judgment, and with Amazon's overhead fees the session cost approximately $670 (U.S.).

In table 1, we show Turker accuracy on the 1299 test items in the All Docs column. The average response accuracy is 76.5%, which is higher than a random baseline. The average accuracy across Turkers is lower, due to a number of Turkers who only did a few HITs and performed poorly. As we look at successively higher thresholds for number of responses, we see increases in accuracy, with the best Turker performing at an accuracy of 90.5%. We also show Turker performance on the candidates drawn from the 181 perfect documents described above (table 1, Perfect Docs). Comparing the results, the Turkers show some improvement (around 2 percentage points) on these documents.

We can aggregate the annotations from multiple Turkers to produce a consensus judgment for each HIT, in order to compensate for poor performers. For example, a majority vote approach counts each Turker judgment as a vote for the correct label for that HIT (candidate gene-mutation relation). The accuracy for a simple majority approach over all documents (83.8%) is substantially

Table 1. Accuracy for overall Turker responses and various subsets

	All Docs	Perfect Docs
Average response	76.5%	79.3
Average Turker	62.0	68.7
Average 10+ Turker	70.7	73.2
Average 100+ Turker	76.0	78.6
Best Turker	90.5	92.4

better than the average response performance (76.5%). Another approach is to use each Turker's control item accuracy to weight their vote, capitalizing on the intuition that the opinion of Turkers who score well on the control items should count for more. However, this performs no better than the simple majority approach (83.7%). More principled approaches include combining labels from different Turkers using probabilistic classifiers. We can treat each Turker response as an observed feature on the candidate item, with the control items viewed as training data. We applied two such frameworks to the data: a Naïve Bayes classifier achieved 84.5% accuracy, while logistic regression performed at 83.2% accuracy.

5   Discussion

Our initial results are promising. We had no difficulty recruiting qualified Turkers and–of particular note–they returned results remarkably quickly (36 hours from release of the HITs). The best Turker achieved a very respectable accuracy of 90.5%, annotating 1144 of 1264 HITs correctly.

Based on a preliminary error analysis, we believe that the reported results underestimate Turker performance. We reviewed 200 cases (ranked by confidence) where Turker aggregate judgments differed from the gold standard. For 50 HITs, there was a mismatch between the gold standard and EMU (v1.0.16) in treatment of SNPs in non-coding regions of genes. In addition, in some 20 HITs, there was a problem with normalization in the gold standard that caused a mismatch. This suggests that overall Turker accuracy may be significantly higher once such discrepancies are resolved, which we plan to do in our follow up experiments.

Feedback from Turkers leads us to believe that the interface provided an effective way to present annotation decisions to reviewers. Because the Turker/ reviewer is presented with a binary decision with the relevant context made salient (by highlighting the gene and mutation candidates in the text), these decisions can be made quite quickly. This particular task turned out to be well-suited to the use of Amazon Mechanical Turk, but we recognize that there may be significant limitations to the crowdsourcing approach for more complex tasks, such as gene-mutation-disease relations, where deep subject matter expertise is needed. We also note that for this experiment, per-abstract costs were significantly higher than curation done by smart undergraduates: $2.50 (U.S.) per abstract done by five Turkers with aggregate accuracy of 85% vs. $0.50 per abstract for undergraduates at an average accuracy of 92%. However, we expect that these costs can be reduced by refinements in distribution of HITs to Turkers and more sophisticated ways of vetting Turkers and aggregating their results.

Finally, we can make a rough estimate of the importance of the biomedical literature as a source of novel findings for human genetic variants. Of the 1770 explicit gene-mutation relations found from the 810 expert-curated abstracts, two thirds (1163) did not appear in OMIM, Swiss-Prot [19] or dbSNP, suggesting that the literature is a rich source of novel gene-mutation relations, even given the 85.

6   Conclusions

Scalable timely capture of gene-mutation-disease information is of critical importance for the rapidly growing field of personalized medicine. The biomedical literature remains a rich source of such information, and validation of relations extracted from the literature is an important step in this process. We have presented a promising initial experiment using crowdsourcing to validate gene-mutation relations assembled from automatically extracted genes and mutations. We were able to recruit high-performing Turkers; they returned results within a day and a half and provided positive feedback on the task and the interface. This suggests that crowdsourced judgments on the validity of candidate biological relations may provide a scalable rapid turn-around approach to obtaining such information.

Acknowledgements. MGK and ED were supported by National Institutes of Health (NIH) Grant 1K22CA143148 to MGK. KL was supported by NIH grant P50MH094267.

References

1. Amberger, J., Bocchini, C.A., Scott, A.F., Hamosh, A.: McKusick's Online Mendelian Inheritance in Man (OMIM). Nucleic Acids Res. 37(Database issue), 793–796 (2009)
2. Sherry, S.T., Ward, M.H., Kholodov, M., Baker, J., Phan, L., Smigielski, E.M., Sirotkin, K.: dbSNP: the NCBI database of genetic variation. Nucleic Acids Res. 29(1), 308–311 (2001)
3. Thorisson, G.A., Lancaster, O., Free, R.C., Hastings, R.K., Sarmah, P., Dash, D., Brahmachari, S.K., Brookes, A.J.: HGVbaseG2P: a central genetic association database. Nucleic Acids Res. 37(Database issue), D797–D802 (2009)
4. Stenson, P.D., Ball, E.V., Howells, K., Phillips, A.D., Mort, M., Cooper, D.N.: The Human Gene Mutation Database: providing a comprehensive central mutation database for molecular diagnostics and personalized genomics. Human Genomics 4(2), 69–72 (2009)
5. Samuels, M.E., Rouleau, G.A.: The case for locus-specific databases. Nat. Rev. Genet. 12(6), 378–379 (2011)

6. Klein, R.J., Zeiss, C., Chew, E.Y., Tsai, J.Y., Sackler, R.S., Haynes, C., Henning, A.K., SanGiovanni, J.P., Mane, S.M., Mayne, S.T., Bracken, M.B., Ferris, F.L., Ott, J., Barnstable, C., Hoh, J.: Complement factor H polymorphism in age-related macular degeneration. Science 308(5720), 385–389 (2005)

7. Denny, J.C., Ritchie, M.D., Basford, M.A., Pulley, J.M., Bastarache, L., Brown-Gentry, K., Wang, D., Masys, D.R., Roden, D.M., Crawford, D.C.: PheWAS: demonstrating the feasibility of a phenome-wide scan to discover gene-disease associations. Bioinformatics 26(9), 1205–1210 (2010)

8. Tatonetti, N.P., Dudley, J.T., Sagreiya, H., Butte, A.J., Altman, R.B.: An integrative method for scoring candidate genes from association studies: application to warfarin dosing. BMC Bioinformatics 11(suppl. 9), S9 (2010)

9. Caporaso, J.G., Baumgartner Jr., W.A., Randolph, D.A., Cohen, K.B., Hunter, L.: MutationFinder: a high-performance system for extracting point mutation mentions from text. Bioinformatics 23(14), 1862–1865 (2007)

10. Doughty, E., Kertesz-Farkas, A., Bodenreider, O., Thompson, G., Adadey, A., Peterson, T., Kann, M.G.: Toward an automatic method for extracting cancer- and other disease-related point mutations from the biomedical literature. Bioinformatics 27(3), 408–415 (2011)

11. Winnenburg, R., Plake, C., Schroeder, M.: Improved mutation tagging with gene identifiers applied to membrane protein stability prediction. BMC Bioinformatics 10(suppl. 8), S3 (2009)

12. Rebholz-Schuhmann, D., Marcel, S., Albert, S., Tolle, R., Casari, G., Kirsch, H.: Automatic extraction of mutations from Medline and cross-validation with OMIM. Nucleic Acids Res. 32(1), 135–142 (2004)

13. Horn, F., Lau, A.L., Cohen, F.E.: Automated extraction of mutation data from the literature: application of MuteXt to G protein-coupled receptors and nuclear hormone receptors. Bioinformatics 20(4), 557–568 (2004)

14. Erdogmus, M., Sezerman, O.U.: Application of automatic mutation-gene pair extraction to diseases. J. Bioinform. Comput. Biol. 5(6), 1261–1275 (2007)

15. Aronson, A.R.: Effective mapping of biomedical text to the UMLS Metathesaurus: the MetaMap program. In: Proc. AMIA Symp., pp. 17–21 (2001)

16. Aberdeen, J., Bayer, S., Yeniterzi, R., Wellner, B., Clark, C., Hanauer, D., Malin, B., Hirschman, L.: The MITRE Identification Scrubber Toolkit: design, training, and assessment. International Journal of Medical Informatics 79(12), 849–859 (2010)

17. Callison-Burch, C., Dredze, M.: Creating speech and language data with Amazon's Mechanical Turk NAACL HLT. In: 2010 Workshop on Creating Speech and Language Data with Amazon's Mechanical Turk. Association for Computational Linguistics, Los Angeles (2010)

18. Yetisgen-Yildiz, M., Solti, I., Xia, F., Halgrim, S.: Preliminary Experiments with Amazon's Mechanical Turk for Annotating Medical Named Entities. In: NAACL HLT 2010 Workshop on Creating Speech and Language Data with Amazon's Mechanical Turk, pp. 180–183. Assn for Comp. Ling, Los Angeles (2010)

19. Boeckmann, B., Bairoch, A., Apweiler, R., Blatter, M.C., Estreicher, A., Gasteiger, E., Martin, M.J., Michoud, K., O'Donovan, C., Phan, I., Pilbout, S., Schneider, M.: The SWISS-PROT protein knowledgebase and its supplement TrEMBL in 2003. Nucleic Acids Res. 31(1), 365–370 (2003)

The Biomedical Translational Research Information System: Clinical Data Integration at the National Institutes of Health

James J. Cimino

Laboratory for Informatics Development
NIH Clinical Center

The National Institutes of Health comprises 27 institutes and centers, many of which conduct clinical research. Although most of the data collection carried out on human subjects at the NIHs main campus in Bethesda, Maryland takes place at the NIH Clinical Center (CC), only a portion of these data are recorded into the hospitals electronic health record. The remainder are distributed into various systems hosted by other institutes or the laboratories of the individual investigators. In addition, a substantial set of data were collected into a previous electronic health record system, retired in 2004. The Biomedical Translational Research Information System (BTRIS) provides a single source for all of these data. Key features of BTRIS include a unifying data model that captures the commonality of data from different sources, without losing their distinctive features, a unifying terminology model that maps high-level research concepts such as disease classes and laboratory results to low-level details such as specific diagnoses and laboratory tests, a self-service feature to allow users to obtain data without an intermediary, and a de-identified query facility to support users asking new questions of old data. Data include vital signs, laboratory and radiographic test results, clinical notes, and diagnoses and problems. BTRIS provides data visualization tools and access to relevant radiographic images. Current expansion is under way to include genomic data, mass spectrometry data, and electrocardiogram tracings. This presentation will focus on architectural design choices, policy and social issues related to data access and sharing, and experience to date.

O. Bodenreider and B. Rance (Eds.): DILS 2012, LNBI 7348, p. 92, 2012.
© Springer-Verlag Berlin Heidelberg 2012

ONCO-i2b2: Improve Patients Selection through Case-Based Information Retrieval Techniques

Daniele Segagni[1], Matteo Gabetta[2], Valentina Tibollo[1], Alberto Zambelli[1], Silvia G. Priori[1], and Riccardo Bellazzi[1,2]

[1] IRCCS Fondazione Salvatore Maugeri, Italy
{daniele.segagni,valentina.tibollo,alberto.zambelli,silvia.priori}@fsm.it
[2] Università degli Studi di Pavia,
Dipartimento di Ingegneria Industriale e dell'Informazione, Italy
{matteo.gabetta,riccardo.bellazzi}@unipv.it

Abstract. The University of Pavia (Italy) and the IRCCS Fondazione Salvatore Maugeri hospital in Pavia have recently started an information technology initiative to support clinical research in oncology called ONCO-i2b2. This project aims at supporting translational research in oncology and exploits the software solutions implemented by the Informatics for Integrating Biology and the Bed-side (i2b2) research center. The ONCO-i2b2 software is designed to integrate the i2b2 infrastructure with the hospital information system, with the pathology unit and with a cancer biobank that manages both plasma and cancer tissue samples. Exploiting the medical concepts related to each patient, we have developed a novel data mining procedure that allows researchers to easily identify patients similar to those found with the i2b2 query tool, so as to increase the number of patients, compared to the patient set directly retrieved by the query. This allows physicians to obtain additional information that can support new insights in the study of tumors.

Keywords: i2b2, oncology, case-based reasoning.

1 Introduction

Most diseases, including cancer, involve a large number and variety of elements that interact via complex networks. Therefore, simply observing the problem from a clinical or biological point of view is not sufficient for treating a disease like cancer. Moreover, effective treatment of disease requires that the clinicians consider the effects of a patient's personal genetic background. Personalized medicine is a proposed approach to develop treatment regimens that take into account each patient's unique genetic profile, allowing the treatment to fit the specific needs of patient sets with different genetic backgrounds. Gathering and integrating data coming from both clinical practice and research settings, together with the possibility of expanding the patients cohort of interest using an innovative case-based reasoning approach, offer physicians the possibility to view the problem in a more complete way.

O. Bodenreider and B. Rance (Eds.): DILS 2012, LNBI 7348, pp. 93–99, 2012.

The University of Pavia (UNIPV) and the IRCCS Fondazione Salvatore Maugeri hospital (FSM) are collaborating on the ONCO-i2b2 project in order to develop a bioinformatics platform designed to integrate clinical and research data to support translational research in oncology.

The next section of this paper presents background information about the data sources integrated into the data warehouse that forms the basis of the ONCO-i2b2 project. The system architecture section describes, through an example, how the system has been designed. The final section outlines the benefits and limitations of ONCO-i2b2.

2   Background

ONCO-i2b2 is based on the software developed by the Informatics for Integrating Biology and the Bedside (i2b2) research center [1]. I2b2 has delivered an open source suite centered on a data warehouse, which is efficiently queried to find sets of interesting patients through a query tool interface.

The ONCO-i2b2 system gathers data from the FSM pathology unit (PU) database and from the hospital biobank, and integrates them with clinical information from the hospital information system (HIS) [2].

Complexity is just one of the problems faced in the integration process we have implemented in order to provide a robust integrated research environment. The system imports data collected during the clinical practice. These data are often not properly structured and may require further extraction steps (e.g., natural language processing tools may be applied to medical reports in order to extract information from clinical narratives). Also integrated in the database are "-omics" data resulting from high-throughput measurement technologies. Although well formatted, such data also pose an integration challenge because of their sheer volume.

Addressing these different challenges of complexity, scope and scale requires a dedicated integration architecture.

3   System Architecture

In this section we will describe the design principles and the implementation of the information technology (IT) architecture that has been deployed at the FSM hospital. As mentioned in the previous section, our system is based on the i2b2 open source software environment, whose main features are briefly summarized below.

3.1   Integration Process

One of the main functionalities of the ONCO-i2b2 project is related to the ability of gathering data about patients and samples, collected during the day-to-day activities of the Oncology I department of the FSM. ONCO-i2b2 also makes these data available for research purposes in an easy, secure and de-identified way. When a patient is hospitalized, he/she is invited to sign an informed consent to make available for research the samples, specimens and data collected for clinical purposes. Specimens obtained surgically are first analyzed by the pathologists of the PU, who may decide to send the specimens exceeding their expertise (together with the signed informed consent) to the laboratory of experimental oncology. The next step consists in the biobank storage of biospecimens. Onco-i2b2 is activated when a biopsy is performed to obtain a detailed diagnosis, and a report is generated. The report contains the cancer diagnosis, including the cancer 'stages' and the size of the tumor. These pieces of information are extracted using a dedicated natural language processing (NLP) module and will be used as concepts for running queries within the i2b2 web client. During this phase the selected samples are de-identified through the use of a new barcode, which does not contain any direct information about the donor. At the same time the system integrates clinical data automatically from the FSM HIS and matches this information to the biobank samples. This information is then stored in the i2b2 Clinical Research Chart (CRC), the star schema data warehouse on which i2b2 is based.

3.2   Accessing the Data

The i2b2 web client is a web-based interface that enables researchers to query and analyze the i2b2 data. The primary advantage of the web client is that no software has to be installed on a user's computer. The query environment is always up to date, reducing maintenance and immediately making available any new features to all users.

The i2b2 web client relies on the definition of a set of hierarchies of concepts, mapped in the concept dimension table of the CRC; such hierarchies are referred to as the i2b2 ontology. Using the query builder interface of the i2b2 web client, researchers can combine these concepts together using and/or logical operators in order to query the data and retrieve patient sets. ONCO-i2b2 ontologies contain concepts from international standards, such as the SNOMED CT terminology, TNM [3] and the ICD9 CM classification system, combined with the specific ontologies required by a specific project for representing concepts related to clinical tests, medications, expression profiles data, biobank repositories and demographics.

The Analysis Tools section of the i2b2 web client has been augmented with two novel plug-ins, which allow users to better examine the information related to a given patient set. One of these plug-ins is called Biobank Info plug-in and retrieves, analyzes and displays information related to the location and tissue type of patients' biospecimens stored in the FSM biobank. Information related to patients' informed consent is also displayed.

The second plug-in, GraphMaker, is more general in purpose. It has been designed with the capability of dynamically creating graphical reports that summarize concepts selected by the user. Graphical result may refer to a user-defined patient set or to all patients contained in the i2b2 CRC.

3.3   Enriching the i2b2 Patient Set

Within the ONCO-i2b2 project, a case- based reasoning procedure has been developed, in order to allow researchers to augmentingenhance the patient selection step process with an information retrieval procedure that uses the whole medical concept space related to a patient set and to identifiyes a group of similar individualspatients. This functionality permits to extendsupports the extension of the original patient set obtained with the i2b2 query tool, with to patients that have some similar concepts.

In this section we will detail the algorithm used to compare patients. At the actual current stagte of the project we are mainly focused on the comparison between patients' clinical data and we limit our analysis to dichotomous conceptsbinary variables, only.

The Unified Medical Language System (UMLS) Metathesaurus [4] is used to model the relationship among i2b2 concepts, in order to create a uniform structure that can be used to compare patients, relying based on a normalized layer.

After a patient set has been retrieved using the i2b2 query tool, our procedure finds all concepts related to patients' observations by means of an array containing UMLS concepts. Each concept is , each one represented by its Concept Unique Identifier (CUI), a code that identifies it withinconcepts in the whole UMLS Metathesaurus, and by a boolean modifier, which indicates if the finding referring to the CUI is asserted or negated. The distance measure computed between cases exploits the semantic similarity between concepts in the UMLS ontology.

Given the concepts array of for each patient of the source patient set (P_S) and the concept array of for the each cases in the i2b2 data warehouse (P_x), the system calculates a similarity score (Sim()) between P_S and P_x as:

$$Sim(P_S, P_x) = w^+ \cdot semDist(P_S^+, P_x^+) + w^- \cdot semDist(P_S^-, P_x^-)$$

where P_S^+ and P_x^- are sub-arrays of P_S and P_x containing the CUIs of the concepts asserted, while P_S^- and P_x^- are the sub-arrays for the negated CUIs. w^+ and w^- are normalization factors. The function $semDist()$ computes the semantic distance between two arrays of CUIs based as on [5]:

$$semDist(P_A, P_B) = \frac{\sum_i min_j(clinDist(P_A i, P_B j))}{size(P_A)}$$

where P_A and P_B are arrays of CUIs, $P_A i$ is the i-th component of P_A and $P_B j$ is the j-th componenent of P_B, and the clinical distance clinDist() between two CUIs is the size of the shortest path joining them in the considered UMLS ontology [6]. Among the different types of relations existing between concepts in the UMLS, we only considered the "isFatherparent" relation, only, as discussed in [6]. In the case of two uncorrelated concepts (no path joining them) the system assigns an arbitrary high (user-defined) clinical distance.

This approach allows physicians and researchers to retrieve similar patients that had a clinical history similar to ones of interest. This allows physicians to obtain additional information that can support new insights in the study of tumors.

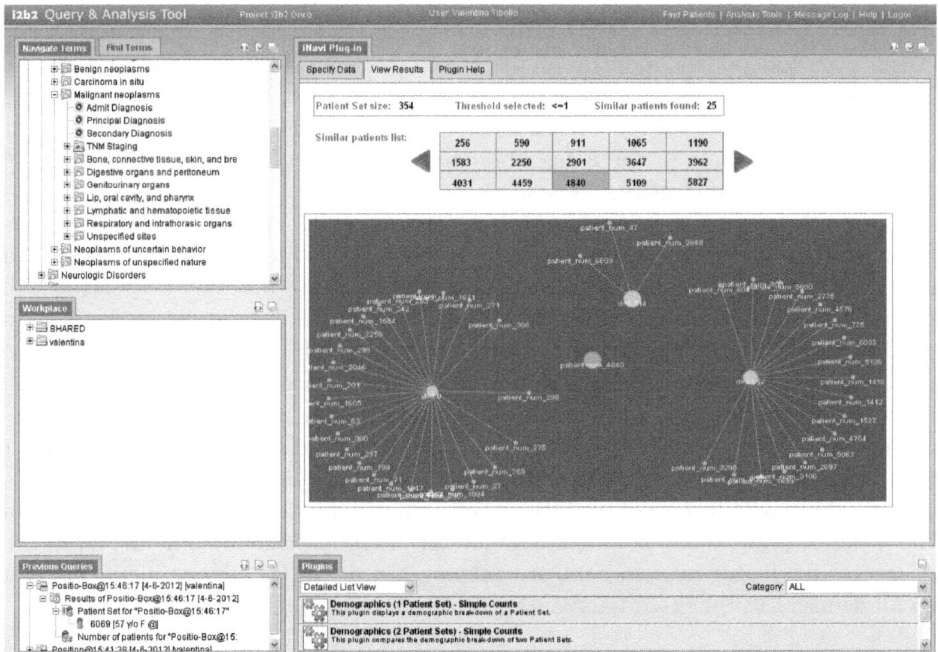

Fig. 1. I2b2 web client with the novel plug-in developed for finding similar patients given a specific patient set. The plug-in shows how the selected new patient (orange cell in the table at the top) is similar to the patients belonging to the patient set. The most similar patients belong to a group of three patients that have a distance d=0.714. The other two groups have d=0.857 (group on the right side) and d=1.0 (group on the left side).

The example showed in Figure 1 represents the results of the method described above. Thanks to a dedicated plug-in, developed for the i2b2 web client interface using the JavaScript InfoVis Toolkit [7], the user is able to:

1. Select a pre-defined patient set that represents the cases related to the medical problem of interest
2. Compute the most similar cases using the ONCO-i2b2 data warehouse (a threshold can be set in order to filter the search)
3. See the results through a graphical visualization that represents the similarity of each new patient to the selected patient set. Patients of the original patient set with the same distance are clustered and represented in a network.

Using this plug-in, a researcher can dynamically find similar patients and explore their related concepts in order to collect useful information potentially related to the medical problem of interest.

4   Results

After a period of few months of tests, the system is up and running at the FSM hospital since December 2010. Currently, the ONCO-i2b2 project consists of 6,713 patients related to breast cancer diagnosis (585 of them have at least one biological sample in the cancer biobank), totaling 47,140 visits and 106,930 observations recorded using 960 concepts.

Although all system functionalities have been fully implemented, the evaluation of the case retrieval procedure has been performed only on a simulated data set focused on a specific medical context. A more detailed evaluation will be performed in the next few months and will involve an analysis of the system performance. It will be carried out over the entire i2b2 data warehouse, before we extend the system by importing information from other hospital wards. To achieve this goal we planned to automatically generate more than 1,000,000 random medical observations, which reflect the concepts distribution in the current system. In this way we can determine where the system needs improvements.

A final remark about computational efficiency. The patient retrieval time is in the order of a few seconds for patient sets up to 1000 patients. The performance decreases for larger patient sets. A cache system is under development in order to allow users to quickly analyze large groups of patients.

5   Conclusions and Future Developments

The ONCO-i2b2 architecture created at FSM is a concrete example of how it is possible to integrate data from different medical areas of interest and to make them available for scientific research. To achieve this goal we had to i) develop "extract, transform and load" (ETL) procedures for automatic population of the i2b2 CRC from PU, biobank and HIS databases with concepts related to cancer research; ii) manage the biospecimen de-identification and the related informed consent; iii) design, develop and configure a case-based reasoning system for patients retrieval; iv) implement new plug-ins to extend the i2b2 web client, in order to allow researchers to better analyze the data collected in the i2b2 CRC.

Exploiting the potential of this IT architecture, the next steps of the project will concern the extension of the data set imported by the HIS, as well as the management of other data from the laboratory tests. We also plan to continue extending the capabilities of the ONCO-i2b2 architecture by improving the case-based reasoning algorithm in order to use numeric and continuous variables coming from clinical practice, laboratory tests or genotyping procedures instead of just binary variables. This will require careful evaluation both in terms of the data representation and storage, and in terms of data security and privacy.

References

1. Murphy, S.N., Weber, G., Mendis, M., Gainer, V., Chueh, H.C., Churchill, S., Kohane, I.: Serving the enterprise and beyond with informatics for integrating biology and the bedside (i2b2). Stud. Health Technol. Inform. 169, 502–506 (2011)
2. Mate, S., Bürkle, T., Köpcke, F., Breil, B., Wullich, B., Dugas, M., Prokosch, H.U., Ganslandt, T.: Populating the i2b2 database with heterogeneous EMR data: a semantic network approach. Stud. Health Technol. Inform. 169, 502–506 (2011)
3. Sobin, L.H., Gospodarowicz, M.K.: Christian Wittekind TNMClassification of Malignant Tumours, 7th edn. Wiley (2009)
4. Xiang, Y., Lu, K., James, S.L., Borlawsky, T.B., Huang, K., Payne, P.R.: k-Neighborhood decentraliza-tion: A comprehensive solution to index the UMLS for large scale knowledge discovery. J. Biomed. Inform., December 2 (2011)
5. Melton, G.B., et al.: Inter-patient distance metrics using SNOMED CT defining relationships. J. Biomed. Inform. 39(6), 697–705 (2006)
6. Caviedes, J., Cimino, J.: Towards the development of a conceptual distance metric for the UMLS. J. Biomed. Inform. 37, 77–85 (2004)
7. InfoVis JavaScipt Toolkit, http://www.thejit.org/

Syntactic-Semantic Frames
for Clinical Cohort Identification Queries

Dina Demner-Fushman and Swapna Abhyankar

National Library of Medicine, Bethesda, MD
{ddemner,abhyankars}@mail.nih.gov

Abstract. Large sets of electronic health record data are increasingly used in retrospective clinical studies and comparative effectiveness research. The desired patient cohort characteristics for such studies are best expressed as free text descriptions. We present a syntactic-semantic approach to structuring these descriptions. We developed the approach on 60 training topics (descriptions) and evaluated it on 35 test topics provided within the 2011 TREC Medical Record evaluation. We evaluated the accuracy of the frames as well as the modifications needed to achieve near perfect precision in identifying the top 10 eligible patients. Our automatic approach accurately captured 34 test descriptions; 25 automatic frames needed no modifications for finding eligible patients. Further evaluations of the overall average retrieval effectiveness showed that frames are not needed for simple descriptions containing one or two key terms. However, our training results suggest that the frames are needed for more complex real-life cohort selection tasks.

1 Introduction

Cohort identification is an essential phase of clinical research and an active area of medical informatics research. Researchers or clinicians first express cohort characteristics (using clinical language familiar to them) as a free text question which subsequently has to be translated into a machine-understandable query to retrieve the relevant information from electronic clinical data warehouses. The descriptions of the cohort inclusion and exclusion criteria are usually complex and multi-faceted. For example, the ClinicalTrials.gov Protocol Registration System allows up to 15,000 characters for the free-text description of the clinical trial eligibility criteria along age, gender and various conditions axes[1]. Traditionally, researchers use formal query languages (directly or with the help of a computer programmer) to query structured clinical data. For example, the Biomedical Translational Research Information System (BTRIS), which is the National Institutes of Health's clinical research data repository, contains pre-defined query templates associated with general retrieval strategies and search filters. Users select the templates relevant to their research question (for example, a lab template for retrieving laboratory test results) and provide the appropriate filter values (such as age, date and specific laboratory test) for the retrieval ([3]).

---

[1] http://prsinfo.clinicaltrials.gov/definitions.html

O. Bodenreider and B. Rance (Eds.): DILS 2012, LNBI 7348, pp. 100–112, 2012.
© Springer-Verlag Berlin Heidelberg 2012

To facilitate direct cohort selection by clinical researchers, Murphy et al. ([15]) have developed a visual approach within the i2b2 hive. The i2b2 visual query tool displays a hierarchical tree of items for the users to choose from in a "Term" panel; the "Query Tool" panels allow users to combine search terms; and the display widgets show the aggregate numbers of patients who match the query criteria. The visual approach was adopted in the Stanford Translational Research Integrated Database Environment that provides a drag and drop cohort discovery tool ([13]). Deshmukh et al. ([5]) found the visual query tools suitable for generating research cohorts based on simple inclusion/exclusion criteria provided that clinical data is structured, coded and can be transformed to fit the logical data models of the i2b2 hive.

Secondary use datasets are becoming more widely available and contain rich collections of both structured and unstructured data. In many such datasets, essential cohort characteristics are only found in the free-text reports; however, efficiently extracting relevant information from narrative text is challenging. Friedman et al. ([6]) presented a natural language processing (NLP) based method for encoding data from clinical text so that the coded data could subsequently be processed by traditional query tools. However, researchers still have to develop the formal queries using templates or visual query tools. Tu et al. ([19]) have developed a semi-automated method for annotating eligibility criteria using the Eligibility Rule Grammar and Ontology (ERGO). The ERGO annotations were then translated to SQL queries to search an electronic health record database for potentially eligible patients. Alternatively, we propose an information retrieval method that takes a researcher's cohort selection criteria expressed in familiar clinical language and automatically extracts the relevant concepts and their relationships into a structured question frame to query narrative reports indexed with a search engine. Similar to translating frames into SQL queries needed for searching relational databases, we automatically translate our frames into a search engine query language. The relations between query concepts are preserved through a set of rules that map frame slots to specific clinical report sections and impose search limits (such as the allowed maximal distance between the terms) on predicates.

We hypothesized that an approach using a domain-specific search engine that considers document structure and question-answering and NLP techniques that incorporate both syntax (i.e., structure) and semantic (i.e., meaning) information would yield robust results. Complex question answering that uses: 1) question classes and named entity classes; 2) syntactic dependency information; and 3) semantic information in the form of predicate-argument structures or semantic frames has been successful in open domain question answering ([16]). In this work, we focused on one question class and combined the syntactic dependency, predicate argument structure and named entities information in a single syntactic-semantic frame for answering cohort selection questions. To fully benefit from the semantic processing of the inclusion criteria, we need to structure the patients' data using an analogous template and unify the patient note frames and the corresponding question frames. In this study, we approximate structuring the patients' narrative data by splitting the clinical documents into sections that correspond to the template slots (such as **past medical history and medications on admission**) and using complex search operators (such as the order of the terms and the distance between them). The Medical Record Retrieval track within the 2011 Text Retrieval Conference (TREC) gave us the opportunity to evaluate this information retrieval method for identifying patient cohorts based on specific inclusion and exclusion criteria ([20]). The cohort descriptions were based on the list of priority topics for comparative effectiveness research issued by the US Institute of Medicine (IOM) of the National Academies ([9]). The proposed query frames could be translated to SQL queries to search over the structured clinical data (if available), as well as to search for eligible patients in free-text reports. The ability to search across both structured and unstructured clinical data will enable complex queries that can identify the most relevant patients.

2   Methods

In this paper, we present the arguably most difficult first step in automatic cohort identification: the automatic generation of question frames from cohort inclusion and exclusion criteria expressed in natural language. We build upon the evidence-based medicine PICO method for asking a well-formed clinical question. Richardson et al. ([17]) first described the PICO method to help clinicians efficiently find the most relevant answers to their clinical questions, and it has been widely incorporated into medical training as part of the evidence-based medicine curriculum. PICO organizes each question into four main parts: 1) **P**atient or **P**roblem; 2) **I**ntervention; 3) **C**omparison intervention (if applicable); and 4) **O**utcome. Previously, we developed an automated method to extract semantic question frames in PICO format for literature-based clinical question answering ([4]). The *frame* is the overall structure that holds all of the relevant concepts for each question, and each frame has four *slots* corresponding to each of the four PICO elements. The system places the relevant concepts from each question into the appropriate PICO slot. In our original work, we modified the PICO format by splitting the **Patient/Problem** slot and adding **Anatomy** to the **Patient** slot, and we merged the **Intervention** and **Comparison** intervention, given that the distinction is not always clear in either the question or in the answer. Inspired by Boxwala et al. ([2]) and Ruiz et al. ([18]), who analyzed query requirements for cohort identification, we further developed our semantic question frame extraction method into a syntactic-semantic method by: 1) refining the basic PICO frame elements with syntactically related words; 2) capturing conjunctions and prepositional phrases; and 3) similar to Jacquemart and Zweigenbaum ([10]), augmenting the basic PICO frame with relational slots that express question elements using predicate-argument structures ([concept]–(relation)–[concept]).

We used 60 training questions created by the second author (SA) to develop our syntactic-semantic method. She based 30 of the training questions on her

Table 1. Syntactic-semantic question frame elements for capturing cohort characteristics

Refined Frame Slot (Basic PICO)	Example of a Refined Frame Slot	Original Question
Age (Patient)	<Age>under50</Age>	patients younger than 50 with hearing loss
Gender (Patient)	<Gender>F</Gender>	women admitted for myocardial infarction who are on hormone-replacement therapy
Population (Patient)	<Population>athletes </Population>	patients seen in the ER with concussion who were athletes
PastMedHx (Problem / Intervention)	<PMH><Prblm> hepatitis</Prblm> <Cause>blood transfusion</Cause></PMH>	patients with a history of hepatitis related to blood transfusion, now with liver cancer
SocialHx (Problem)	<SocialHx>smoking</SocialHx>	patients with a history of smoking as well as personal and family history of lung cancer
AdmitProblem (Problem)	<AdmitPrblm>stroke </AdmitPrblm>	patients admitted for stroke who arrived too late for tPA administration
DischargeProblem (Problem)	<DischPrblm>wound infection</DischPrblm>	patients who developed a wound infection during the current hospital stay
Problem (Problem)	<Prblm>concussion</Prblm>	patients seen in the ER with concussion who were athletes
Finding (Problem)	<Finding>hearing loss</Finding>	patients younger than 50 with hearing loss
Complications_Of (Problem)	<ComplicationsOf> <Prblm>pneumothorax</Prblm> <Cause>VATS</Cause> </ComplicationsOf>	patients who developed a pneumothorax as a complication of VATS
Allergies (Problem)	<Allergies>drug allergy</Allergies>	patients with a known drug allergy who received a drug in the same allergy class
Anatomy (Patient)	<Anatomy>cervical spine</Anatomy>	patients admitted for surgery of the cervical spine for fusion or discectomy
MedBeforeAdm (Problem / Intervention)	<MedBeforeAdm><Drug> </Drug> <Prblm> osteoporosis OR osteopenia</Prblm> </MedBeforeAdm>	women with hip or vertebral fracture despite being on medication for osteoporosis or osteopenia
MedOnDisch (Problem / Intervention)	<MedOnDisch><Drug> <NEG>inhaled steroids </NEG> </Drug> <Prblm> COPD </Prblm> </MedOnDisch>	patients with COPD who were not discharged on inhaled steroids
MedForProblem (Problem/ Intervention)	<MedForPrblm><Drug> ritalin</Drug><Prblm> depression</Prblm> </MedForPrblm>	patients on Ritalin for depression
ProcBeforeAdm (Intervention)	<ProcBeforeAdm>dialysis </ProcBeforeAdm>	patients admitted for complications due to renal failure despite being on dialysis
ProcForProblem (Problem / Intervention)	<ProcForPrblm> <Proc>ablation</Proc> <Prblm>atrial fibrillation </Prblm> </ProcForPrblm>	patients with atrial fibrillation treated with ablation
Procedure (Intervention)	<Procedure>surgery<MOD> robotic-assisted </MOD></Procedure>	patients who had robotic-assisted surgery
FamilyHx (Problem /Intervention)	<FamilyHx><Prblm>lung cancer</Prblm></FamilyHx>	patients with a history of smoking as well as personal and family history of lung cancer
DischDest (Outcome)	<DischDest>skilled nursing facility</DischDest>	patients with dementia who were discharged to a skilled nursing facility or other institutional setting
Encounter Location ()	<Location>ER</Location>	patients seen in the ER for low back pain who were not admitted to the hospital

patient encounters and on interesting topics in recent issues of the General Medicine Journal Watch ([11]), and the other 30 on the IOM priority topics (the Medical Records Retrieval track organizers told the track participants in advance that the test topics would be based on the IOM priority topics and did not restrict access to those topics during the development period). Together we expanded the four basic PICO slots into more than twenty refined slots; for example, the original *Patient* slot was split into *Age, Gender, Population,* and *Anatomy.* Both authors then manually encoded 30 training questions each using the refined frame slots, and subsequently they reviewed all 60 training questions and finalized the refined frames together. We developed frames capable of capturing nuances of the question, such as temporal relations and specific groups of patients. For example, we defined three medication slots (*medications before admission, on discharge,* and the fallback, *medication for problem*). These distinctions are needed to encode (and answer) temporal questions, such as *Find patients with HIV admitted for a secondary infection who were not on prophylaxis for opportunistic infection.* Table 1 presents the final set of the frame slots. Note that we chose the surface representation of our frame slots in XML format for the convenience of then automatically translating the frames to the query syntax of the search engines, Lucene[2] and Essie ([8]), used for retrieval. Both search engines rank results according to the likelihood of their relevance and provide complex syntax that allows the user to structure queries beyond simple keyword searches.

2.1   Frame Extraction System

Our automatic system extracts the frames in four steps. In the first step, the system submits the question to MetaMap ([1]) with the default settings to extract the Unified Medical Language System® (UMLS®) ([12]) concepts. For each concept, the system stores the lexical match with offset and length, negation status and semantic types in a lookup table.

In the second step, the system uses regular expressions to extract patient demographics and social history. The patterns for age include a small vocabulary of age-related terms (for example, preemie, toddler, tween) and a library of regular expressions for identifying specific ages and age ranges (for example, $(\backslash d+)\backslash W^*years\backslash W^*of\backslash W^*age$). Our entire gender look-up list is very small (female, girl, gravida, her, lady, ladies, she, woman, women, boy, he, his, male, man, men, gentleman, gentlemen). The Population slot is currently limited to occupations and ethnicities defined by the UMLS semantic types Professional or Occupational Group and Population Group, respectively. The patterns for social history are currently limited to identifying smoker status, alcohol consumption and illicit drug use. We used lexico-semantic patterns to extract the Complications_of relation. Our patterns combine semantic categories and lexemes, for example, we created the [*concept_Problem*]*s/p*[*concept_any*|*word_noun*] pattern based on the training question *patients admitted for injuries s/p fall.*

---

[2] http://lucene.apache.org/core/

Table 2. Rules and examples for the syntactic-semantic question frames

Refined Frame slot	Rules and constraints	Example
UMLS concept augmented with modifiers	If dependency ∈ modifier & governor ST ∈ (Problem| Intervention)⇒add modifiers to the term	Q:...MRSA endocarditits... endocarditis amod MRSA & endocarditis[dsyn]⇒<Prblm> endocarditis<MOD>MRSA </MOD></Prblm>
Conjunction	If dependency ∈ (AND|OR) & governor | dependent ∈ (Problem|Intervention)⇒ join terms	Q: ...staging or monitoring of cancer... staging conj_or monitoring & monitoring[hlca]⇒ <Proc>staging OR monitoring</Proc>
Admit_problem	If dependency ∈ (prep_with| prep_for) & governor = admit & dependent ∈ Problem⇒admit_problem	Q: ...admitted with an asthma exacerbation admitted prep_with asthma exacerbation & asthma exacerbation[fndg]⇒ <AdmitProblem>asthma exacerbation </AdmitProblem>
Med_for_problem Proc_for_problem	If dependency path contains a treatment indicator, Intervention and Problem ⇒ if Intervention ∈ Drug ⇒ Med_for_problem, else ⇒ Proc_for_problem	Q: ...monoclonal antibody treatment for inflammatory bowel disease treatment amod monoclonal antibody treatment prep_for inflammatory bowel disease & inflammatory bowel disease[dsyn] monoclonal antibody[aapp,imft] ⇒ <MedForPrblm><Drug>monoclonal antibody</Drug><Prblm> inflammatory bowel disease</Prblm></MedForPrblm>

In the third step, the system processes the question sentences using the Stanford dependency parser ([14]). To prevent the parser from breaking-up multi-word concepts, the system first concatenates all of the words in a concept. We focused on extracting a limited set of typed dependency relations, conjunctions, and modifiers. The system only populates the frame slot if the semantic and syntactic constraints are satisfied. Table 2 shows the examples of the rules. If a rule is applied, the concepts used in that rule are marked as "used" in the look-up table.

After completing iterations over the dependency paths, in the final (fourth) step, all of the remaining unused concepts that could populate the four basic PICO slots are added to the frame. That is, if the lookup table for the question concepts contains concepts in the semantic groups **Disorders (Problems)**, **Interventions** or **Anatomy** that are not already marked as used, the concepts populate the traditional PICO frame slots. Although every question passes through the four modules that implement the four steps, the question frame may pass through a module without changes if none of the rules or patterns applies. Table 3 illustrates each step of the automatic frame extraction for one of the test questions.

2.2   Evaluation

We conducted both an intrinsic and an extrinsic evaluation of the test frames automatically generated from the TREC Medical Record Retrieval track test topics. Since DDF implemented the frame extraction system, only SA conducted

Table 3. Four question-processing steps

Step	Process	Result
0	Original TREC test question	Patients with complicated GERD who receive endoscopy
1	UMLS concept extraction via MetaMap 2010	Disease or Syndrome: GERD Diagnostic Procedure: endoscopy *Explanation*: "GERD" and "endoscopy" are identified as relevant concepts and assigned the semantic types "Disease or Syndrome" and "Diagnostic Procedure," respectively
2	Using regular expressions to extract demographics and social history	n/a for this question
3	Extracting dependencies using the Stanford Dependency Parser	Modifier: complicated Proc_for_problem: endoscopy|GERD *Explanation*: the word "complicated" is identified as a modifier of "GERD" and the relationship between "GERD" and "endoscopy" is identified as procedure for problem
4	Assign concepts that have not yet been used to the four basic PICO slots	n/a for this question

Final frame result: <ProcForPrblm><Proc>endoscopy</Proc>
<Prob>GERD<MOD>complicated</MOD></Prob></ProcForPrblm>

the intrinsic evaluation to judge the accuracy of the automatic frame extraction. She judged a frame to be correct if the system extracted all of the test question elements and each one was placed into its correct slot.

Even the perfectly generated frames are not guaranteed to find eligible patients. There are three possible reasons for failure: 1) there are no eligible patients in the dataset; 2) the algorithm for converting question frames to complex search engine queries is incorrect; and 3) the query needs to be more complex than correctly combining the key terms provided in the description of the inclusion criteria in complex searches. In our extrinsic evaluation, we focused on the third reason: the complexity of the query and the modifications to the terms, template and complex searches needed for near prefect precision at top ten potential cohort participants.

For the extrinsic evaluation, we used the Essie corpus analysis and mining tool ([8]) and the Medical Record Retrieval track document collection. Essie is a domain specific search engine with built-in UMLS-based synonymy expansion that developers at the National Library of Medicine created to support NLM's ClinicalTrials.gov. Essie has been used for that purpose since 2001. The TREC document collection contained over 100,000 reports from over 17,000 patient visits. Each visit was associated with one or more reports; for example, if the visit was to the `Emergency Department (ED)`, only the `ED` note was associated with that visit, but if the visit was a multiple-week hospital stay, dozens of documents of various types (e.g., progress notes, radiology reports, discharge summary) were associated with that single visit. For every test question, each author independently used the Essie interface to manually generate a query, reviewed the top ten patients' visits that Essie returned, and refined the query until she had created the ideal question to return the most relevant visits. Both authors then compared the manual queries developed using Essie to the automatic frames to evaluate the differences between the manual and automatic queries. We also compared the visits returned by each method to see how the query differences impacted the relevance of the visits that were retrieved. Here, we only evaluate how many automatic frames needed modifications and the nature of the modifications.

Finally, we compared the average performance of the baseline queries to that of the frame-based queries. For the baseline queries, the original free-text descriptions of the inclusion criteria were submitted to a search engine without any modifications. Clinical documents are not likely to contain exact matches of these long descriptions, therefore, we allow the search engine to arbitrarily break the baseline queries into phrases and remove low-frequency query terms ("lossy expansion").

3   Results

In the intrinsic evaluation, SA evaluated the accuracy of the automatically generated frames created from the test questions provided by the TREC Medical Records track organizers. Out of the 35 frames 34 were accurate and one was incorrect. The reason for the error in the frame (shown in Table 4) is the lack of the appropriate cue in our pattern set for the Complications_of slot. Once the *secondary to* cue was added to the patterns (in a system that was not used in the TREC evaluation), the correct frame was extracted.

In the extrinsic evaluation, we evaluated the usefulness of the automatic frames for cohort identification. We could evaluate only 34 frames because one

Table 4. Automatic and manually corrected question frames for the question "Adult patients who presented to the emergency room with with anion gap acidosis secondary to insulin dependent diabetes"

Automatically generated frame	Correct frame
<Age>adult</Age>	<Age>adult</Age>
<Prblm>insulin dependent diabetes <MOD> secondary </MOD></Prblm> <Prblm>anion gap acidosis</Poblm> <Location>emergency room</Location>	<ComplicationsOf> <Prblm> anion gap acidosis </Prblm> <Cause> insulin dependent diabetes </Cause> </ComplicationsOf> <Location>emergency room</Location>

Table 5. Modifications needed for near perfect precision. Questions for which the original frames retrieved few relevant documents in the top ten are marked with an asterisk.

	Test question	Automatic frame	Modifications [Modification type]
1	Hospitalized patients treated for methicillin-resistant Staphylococcus aureus (MRSA) endocarditis	`<Prblm>`endocarditis `<MOD>`MRSA `</MOD></Prblm>`	Prblm: (SBE OR endocarditis) AND (staph OR MRSA) [Domain knowledge of the disease]
2*	Patients with ductal carcinoma in situ (DCIS)	`<Prblm>`ductal carcinoma`</Prblm>` `<Prblm>`DCIS `</Prblm>`	Prblm: Ductal carcinoma in situ OR (breast cancer AND in situ) [Domain knowledge of the disease]
3*	Patients treated for vascular claudication surgically	`<ProcForPrblm>` `<Proc>`surgically `</Proc>` `<Prblm>`claudication `<MOD>` vascular`</MOD>` `</Prblm>` `</ProcForPrblm>`	Prblm: vascular claudication OR ("peripheral vascular disease" AND calf) Proc: endarterectomy OR popliteal [Domain knowledge of the disease and the procedures]
4*	Patients with chronic back pain who receive an intraspinal pain-medicine pump	`<ProcForPrblm>` `<Proc>`pump`<MOD>`intraspinal`</MOD>` `<MOD>`pain-medicine`</MOD>` `</Proc>` `<Prblm>` chronic back pain `</Prblm>` `</ProcForPrblm>`	Drug: (Intrathecal OR subarachnoid OR intraspinal OR epidural) AND ("morphine pump" OR "intrathecal pump" OR "pain pump" OR "dilaudid pump" OR "opioid pump" OR "epidural pain") [Domain knowledge of pain medications and administration routes]
5	Adult patients who presented to the emergency room with with anion gap acidosis secondary to insulin dependent diabetes	See Table 4	Prblm: anion gap acidosis OR DKA OR ketoacidosis OR metabolic acidosis [Domain knowledge of the disease]
6*	Patients admitted for hip or knee surgery who were treated with anticoagulant medications post-op	`<AdmitProblem>` hip OR knee surgery `</AdmitProblem>` `<Drug>`anti-coagulant `</Drug>`	Drug: heparin OR warfarin OR Clopidogrel OR Ticlopidine OR Enoxaparin OR anticoagulants [Domain knowledge of medications]
7*	Patients who underwent minimally invasive abdominal surgery	`<Procedure>` abdominal surgery`<MOD>`invasive `</MOD>` `<MOD>`minimally `</MOD>` `</Procedure>`	Proc: (Laparoscopic OR minimally invasive OR MIS) NEAR (abdominal OR bariatric Or gastrojejunostomy OR appendectomy OR colectomy OR sigmoidectomy OR cholecystectomy) [Domain knowledge of procedures]
8*	Patients admitted for care who take herbal products for osteoarthritis	`<MedForPrblm>` `<Drug>`herbal products`</Drug>``<Prblm>` osteoarthritis `</Poblm>` `</MedForPrblm>`	Drug: Capsaicin OR capzasin OR arthritis formula OR soy OR Boswellia OR licorice OR cohosh OR hawthorn OR castor OR "cranberry capsule" OR "cranberry tablet" OR echinacea OR Glucosamine OR ubiquinone OR thistle OR Gingko-Biloba OR primrose OR aloe OR cinnamon OR flaxseed [Domain knowledge of medications]
9	Patients admitted with chronic seizure disorder to control seizure activity	`<Problem>` seizure disorder `<MOD>`chronic `</MOD>` `</Problem>`	Prblm: seizure disorder OR "status epilepticus" [Domain knowledge of the disease]

of the original cohort descriptions had no relevant documents in the collection. Based on manual review of how relevant the top ten visits retrieved by the automatic frames were to the cohort criteria, we found that 25 test question frames did not need any modifications, and only nine test question frames did. Of those nine, six would have failed to find most relevant documents without modifications. For the remaining three questions, modifications targeted recall and improving precision. In all cases, modifications required domain knowledge beyond the UMLS synonymy. For the questions that would have failed, the drug classes or high level descriptions of the procedures needed to be expanded with specific instances. See for example, the expansion for *herbal products* (question 8 in Table 5). In the comparison of the overall average performance, the frame-based queries did not provide the anticipated advantage compared to the baseline queries.

4   Discussion

Formally representing the essence of the cohort characteristics for comparative effectiveness studies can potentially streamline the cohort identification process. Previous analysis of the basic PICO frames, which were designed to formally represent clinical questions, showed that the framework is best suited for representing therapy questions and considerably less suitable for diagnosis, etiology, and prognosis questions. The basic framework cannot capture the fine-grained relationships between frame elements, or model temporal/state information and anatomical relations ([7]), which are exactly the elements needed to accurately capture the study cohort characteristics. We hypothesized that expanded syntactic-semantic PICO frames would compensate for the shortcomings of the basic PICO frames, potentially at the cost of being more brittle.

One benefit of our method is that the syntactic-semantic query frame is generated completely independently of the patient records to be queried, so the same frame can be used to search multiple disparate databases, regardless of the database structure. Different database sources might process patient notes in different ways, either by applying a frame structure similar to the query frames, encoding the data using NLP as described by Friedman et al ([6]), dividing the note into sections as we did for TREC 2011, or taking the text of the note without any modification. Given that the query frame is independent of the clinical note structure, the same query frame can be matched with different local note structures, which is key for data integration from multiple sites. The common query frame method can be used to query data from different providers and hospital systems not only for cohort identification, but also for assessing quality metrics and for health information exchange.

Our first concern in developing the frames was determining the minimal set of slots capable of capturing all necessary fine-grained details. For example, we define medications on discharge to capture medications administered only during the hospital encounter and distinguish those from medications on admission and take-home medications, but we have only one slot for procedures (to capture procedures performed during the hospital encounter). We assume that procedures performed before the current encounter are associated with past medical history, and all procedures not associated with the past medical history occurred during the encounter. Only 10 of our 21 slots were needed to encode the TREC test questions: `Age`, `Gender`, `Anatomy`, `Complications_of`, `AdmitProblem`, `Finding`, `Problem`, `Procedure`, `ProcForPrblm`, and `MedForPrblm`. We thoroughly verified that the basic frame slots (used more extensively in the test questions than in the training questions) were appropriate for capturing the question information. Indeed, the test questions seemed less complex than our training set, and if they were used for actual cohort selection, it is likely that more patients would have to initially be screened and then excluded in subsequent selection steps. For example, eight questions had the disorder as the only selection criterion. This seeming simplicity of the test questions is, however, understandable. Traditionally, the first year of a TREC track (as this was) gives the participants an opportunity to focus on the document set and accomplishing the test task within a short timeframe. The complexity of the task increases in the subsequent evaluations. Future work on more complex questions with a larger set of criteria for retrospective study cohorts will determine if our current set of 21 slots is capable of capturing all of the necessary details and if the syntactic-semantic frames will provide the hypothesized advantages over the baseline queries.

Our second concern was the potential brittleness of the approach that relies heavily on the typed dependency parse tree. In this evaluation, all syntactic-semantic extraction rules that fired during the extraction of the test question frames were triggered correctly and populated the correct frame slots.

The limitation of our study is that the relatively small number of the syntactic-semantic extraction rules was developed using a relatively small set of training questions. We need to test if the number of slots and rules can be kept at a manageable size with a larger set of questions. This opportunity will present itself in the TREC 2012 Medical Records evaluations. Note, that in the subsequent evaluations we do not have to rely solely on translating the frames to the search engine query languages. Instead, we could apply the proposed method (alone or after an initial search step) to the patients' notes. Then, frame unification or a constrained frame matching approach could be applied to both the query and the patients' cases frames. In fact, we tested the constrained frame matching approach in answering clinical questions with extracts from MEDLINE® citations and found it to be more accurate than the information retrieval approaches alone ([4]). We anticipate similar results could be achieved for the cohort identification task. Generation of the patients' cases frames, however, will require additional research.

5 Conclusion

The secondary use of clinical data for cost- and comparative- effectiveness studies is a burgeoning area of clinical research. An automatic system capable of identifying patients based on a textual description of the inclusion and exclusion criteria will potentially speed up the process of cohort identification, as well as enable exchange of health information and evaluation of clinical quality metrics. Our evaluation of capturing the inclusion/exclusion criteria of a comparative-effectiveness study cohort expressed as a natural language question shows that syntactic-semantic frames can accurately capture the desirable patients' characteristics. Further evaluation of the frames' retrieval effectiveness showed that a third of the frames needed modifications because of the mismatch between the high-level descriptions of the criteria in the question and the more specific terms that define the diagnoses, procedures, and medications in the clinical records. Future work will involve further evaluation of the overall average retrieval effectiveness of the automatic syntactic-semantic frames and refining the frame extraction system.

Acknowledgments. This research was supported in part by the Intramural Research Program of the National Institutes of Health (NIH), National Library of Medicine (NLM).

References

1. Aronson, A.R., Lang, F.M.: An overview of MetaMap: historical perspective and recent advances. J. Am. Med. Inform. Assoc. 17(3), 229–236 (2010)
2. Boxwala, A., Kim, H., Choi, J., Ohno-Machado, L.: Understanding data and query requirements for cohort identification in clinical research studies. In: AMIA Annu. Symp. Proc., p. 95 (2011)
3. Cimino, J.J., Ayres, E.J.: The clinical research data repository of the US National Institutes of Health. Stud. Health Technol. Inform. 160(Pt 2), 1299–1303 (2010)
4. Demner-Fushman, D., Lin, J.: Answering Clinical Questions with Knowledge-Based and Statistical Techniques. Computational Linguistics 33(1), 63–103 (2007)
5. Deshmukh, V.G., Meystre, S.M., Mitchell, J.A.: Evaluating the informatics for integrating biology and the bedside system for clinical research. BMC Med. Res. Methodol. 9, 70 (2009)
6. Friedman, C., Shagina, L., Lussier, Y., Hripcsak, G.: Automated encoding of clinical documents based on natural language processing. J. Am. Med. Inform. Assoc. 11(5), 392–402 (2004)
7. Huang, X., Lin, J., Demner-Fushman, D.: Evaluation of PICO as a knowledge representation for clinical questions. In: AMIA Annu. Symp. Proc., pp. 359–363 (2006)
8. Ide, N.C., Loane, R.F., Demner-Fushman, D.: Essie: a concept-based search engine for structured biomedical text. J. Am. Med. Inform. Assoc. 14(3), 253–263 (2007)
9. Institute of Medicine of the National Academies (IOM): 100 initial priority topics for comparative effectiveness research, http://www.iom.edu/Reports/2009/ComparativeEffectivenessResearchPriorities.aspx
10. Jacquemart, P., Zweigenbaum, P.: Towards a medical question-answering system: a feasibility study. Stud. Health Technol. Inform. 95, 463–468 (2003)
11. JournalWATCH® General Medicine: http://general-medicine.jwatch.org/ (updated August 16, 2011, accessed August 16, 2011)
12. Lindberg, D.A.B., Humphreys, B.L., McCray, A.T.: The Unified Medical Language System. Meth. Inform. Med. 32, 281–291 (1993)
13. Lowe, H.J., Ferris, T.A., Hernandez, P.M., Weber, S.C.: STRIDE–An integrated standards-based translational research informatics platform. In: AMIA Annu. Symp. Proc., pp. 391–395 (2009)
14. de Marneffe, M.-C., MacCartney, B., Manning, C.D.: Generating Typed Dependency Parses from Phrase Structure Parses. In: LREC 2006 (2006), http://nlp.stanford.edu/pubs/LREC06_dependencies.pdf (accessed August 16, 2011)

15. Murphy, S.N., Barnett, G.O., Chueh, H.C.: Visual query tool for finding patient cohorts from a clinical data warehouse of the Partners HealthCare system. In: Proc. AMIA Symp., p. 1174 (2000)
16. Narayanan, S., Harabagiu, S.: Question Answering based on Semantic Structures. In: International Conference on Computational Linguistics COLING 2004, Geneva, Switzerland (2004)
17. Richardson, W.S., Wilson, M.C., Nishikawa, J., Hayward, R.S.: The well-built clinical question: a key to evidence-based decisions. ACP J. Club 123, A12–3 (1995)
18. Ruiz, E.E., Chilov, M., Johnson, S.B., Mendonça E.A.: Developing multilevel search filters for clinical questions represented as conceptual graphs. In: AMIA Annu. Symp. Proc., p. 1118 (2008)
19. Tu, S., Peleg, M., Carini, S., Bobak, M., Ross, J., Rubin, D., Sim, I.: A practical method for transforming free-text eligibility criteria into computable criteria. J. Biomed. Inform. 44(2), 239–250 (2011)
20. Voorhees, E., Tong, R.: Overview of the TREC 2011 Medical Records Track. In: The Twentieth Text REtrieval Conference Proceedings TREC 2011, Gaithersburg, MD. National Institute for Standards and Technology (2011)

Author Index

Aberdeen, John   83
Abhyankar, Swapna   100

Barker, Ken   53
Bayer, Sam   83
Bellazzi, Riccardo   93
Benik, Joseph   21
Bradley, Ray M.   56
Buendia, Patricia   56
Burger, John D.   83

Chang, Caren   21
Cimino, James J.   92

Demner-Fushman, Dina   100
Doughty, Emily   83

Embury, Suzanne M.   37

Gabetta, Matteo   93

Hirschman, Lynette   83
Hussels, Philipp   5

Jean-Mary, Yves R.   56

Kabuka, Mansur R.   56
Kann, Maricel G.   83

Lee, Kyungjoon   83
Leser, Ulf   5

Maskat, Ruhaila   37

Ogbuji, Chimezie   71
Ostell, James M.   1

Palma, Guillermo   21
Paton, Norman W.   37
Priori, Silvia G.   93

Raschid, Louiqa   21

Segagni, Daniele   93
Shironoshita, E. Patrick   56

Thor, Andreas   21
Tibollo, Valentina   93
Tresner-Kirsch, David   83
Trißl, Silke   5

Vidal, Maria-Esther   21

Wellner, Ben   83

Xu, Rong   71

Zambelli, Alberto   93